Models of Discovery and Creativity

Joke Meheus • Thomas Nickles
Editors

Models of Discovery and Creativity

Springer

Editors
Joke Meheus
Universiteit Gent
Vakgroep Wijsbegeerte
en Moraalwetenschap
Blandijnberg 2
9000 Gent
Belgium
Joke.Meheus@UGent.be

Thomas Nickles
Department of Philosophy
University of Nevada, Reno
Reno, NV 89557
USA

ISBN 978-90-481-3420-5 e-ISBN 978-90-481-3421-2
DOI 10.1007/978-90-481-3421-2
Springer Dordrecht Heidelberg London New York

Library of Congress Control Number: 2009937610

Printed on acid-free paper

Springer is part of Springer Science+Business Media (www.springer.com)

Contents

Foreword vii

Preface ix

Unexpected discoveries, Graded Structures, and the Difference between Acceptance and Neglect 1
Hanne Andersen
1 The Conceptual Analysis 3
2 Nuclear Physics 4
3 Philosophical Morals 22

Conceptual Comparison and Conceptual Innovation 29
Harold I. Brown

Discovering Mechanisms in Molecular Biology Finding and Fixing Incompleteness and Incorrectness 43
Lindley Darden
1 Introduction 43
2 Characterization of Mechanisms 45
3 Revision of Incomplete Schemata 47
4 Revision of Incorrect Schemata 50
5 Conclusion 53

On the Role of Thought-Experiments in Mathematical Discovery 57
Eduard Glas
1 Archimedes's Method 58
2 Impossible Numbers 60
3 Conclusion 63

Experimental Systems, Investigative Pathways, and the Nature of Discovery 65
Frederic L. Holmes

Abduction as a Heuristic Constraint 81
Scott A. Kleiner
1 Introduction 81
2 The Problem of Abduction 83
3 Evolutionary Biology 86

4 Conclusions 92

Creative Abduction and Hypothesis Withdrawal 95
Lorenzo Magnani
1 Change in Theoretical Systems 95
2 Abduction: Sentential, Model-Based, Manipulative 97
3 Governing Inconsistencies in Abductive Reasoning 104
4 Withdrawing Unfalsifiable Hypotheses 114

Conceptual Change: Creativity, Cognition, and Culture 127
Nancy J. Nersessian
1 Introduction 127
2 Interpreting Conceptual Practices: Cognitive-Historical Analysis 127
3 Cognition and Culture: Situated and Distributed Cognition 131
4 Creativity in Conceptual Change: The Role of Model-Based Reasoning 137
5 Model-based Reasoning as Situated and Distributed Reasoning 153
6 Culture and Cognition: Implications for Creativity 158

The Strange Story of Scientific Method 167
Thomas Nickles
1 Introduction 167
2 Traditional Views of Method and Discovery 169
3 Scientific Method (So-Conceived) Is Impossible 171
4 Reasons for Optimism? 181
5 Two Objections 184
6 The Triumph of the Darwinian Method? 186
7 BV+SR: Madness or Method? 190
8 The Generality Question and the NFL Theorems 198
9 The Classical Discovery Program Revisited 200

Tradition and Innovation: Exploring and Transforming Conceptual Structures 209
Matti Sintonen
1 Introduction 209
2 Taditionalists and Iconoclasts 210
3 Scientific Structures 212
4 Applied and Intractable Fields 214
5 Discovery in the Mature Sciences 216
6 Exploring Paradigms 218

A Purposeful Alliance in the Service of Creative Research
The Network of Vitamin Investigators 223
Petra Werner
1 Introduction 223
2 The Significance of Collective Work 224
3 How are the Results Evaluated from the Current Perspective? 226
4 How Effective was the Network? 234
5 Conclusion 235

Index 237

Foreword

Since the origin of the modern sciences, our views on discovery and creativity had a remarkable history. Originally, discovery was seen as an integral part of methodology and the logic of discovery as algorithmic or nearly algorithmic. During the nineteenth century, conceptions in line with romanticism led to the famous opposition between the context of discovery and the context of justification, culminating in a view that banned discovery from methodology. The revival of the methodological investigation of discovery, which started some thirty years ago, derived its major impetus from historical and sociological studies of the sciences and from developments within cognitive psychology and artificial intelligence.

Today, a large majority of philosophers of science agrees that the classical conception as well as the romantic conception are mistaken. Against the classical conception, it is generally accepted that truly novel discoveries are not the result of simply applying some standardized procedure. Against the romantic conception, it is rejected that discoveries are produced by unstructured flashes of insight.

An especially important result of the contemporary study concerns the availability of (descriptive and normative) models for explaining discoveries and creative processes. Descriptive models mainly aim at explaining the origin of novel products; normative models moreover address the question how rational researchers should proceed when confronted with problems for which a standard procedure is missing.

The present book provides an overview of these models and of the important changes they induced within methodology. As appears from several papers, the methodological study of discovery and creativity led to profound changes in our conceptions of justification and acceptance, of rationality, of scientific change, and of conceptual change.

The book contains contributions from both historians and philosophers of science. All of them, however, are methodological in the contemporary sense of the term. The central values of this methodology are empirical accurateness, clarity and precision, and rationality. The different contributions realize these values by their interdisciplinary nature. Some philosophically oriented papers rely on historical case studies and results from the cognitive sciences, others on recent results from the computer sciences and/or non-standard logics. The historically oriented papers address central philosophical questions and hypotheses.

Acknowledgments

The editors are indebted to the Research Foundation – Flanders and to Gitte Callaert, Bert Leuridan, Friederike Schröder-Pander, and Stephan van der Waart van Gulik for their help in preparing the manuscript.

JOKE MEHEUS

Preface

At the end of October 1978, I had the privilege of organizing a conference on scientific discovery at the University of Nevada, Reno, USA. That was the first Guy Leonard (Memorial) Conference at UNR. Sam Goudsmit, co-discoverer of electron spin, then a professor at UNR after a distinguished career at Michigan, gave the opening lecture, "Physics in the Twenties", just a few days before his own death. The conference included around fifty participants from six countries, and the proceedings were eventually published by Reidel in two volumes. Herbert Simon and others working in artificial intelligence and neighboring fields had for some years focused on discovery and problem solving, but the Reno conference is often credited with helping to legitimize the topic for philosophers of science, epistemologists, and even some logicians.

As that conference ended, Lindley Darden remarked that it would be nice to assemble a similar group twenty years hence to determine what progress had been made. As it turns out, it was exactly twenty years later that Joke Meheus at Ghent University organized the conference to which the volume you are holding is devoted. The logic group at Ghent, headed by Diderik Batens, had by then devoted many years of research to developing logics that better capture the way in which people actually think in problem-solving contexts. They were, and have continued to be, among the most important "friends of discovery". By now work is well underway in many quarters on various philosophical or logical aspects of discovery, understood in a broad enough sense to include construction of novel models and research programs. Some of the most impressive work is being accomplished by Clark Glymour's group in the Philosophy Department at Carnegie-Mellon University in Pittsburgh, e.g., work on formal learning theory and on causal Bayes networks. Others, myself included, are taking a more historical approach.

Although Joke Meheus insisted that I be listed as a co-editor of this volume, I must confess that the International Congress on Discovery and Creativity was hers and Diderik's idea and that she deserves all praise for organizing the conference and for editing this volume. A great deal of effort was involved. The Ghent congress was on the same size scale as the one in Reno but, thanks in part to its more convenient location, more international. Joke even arranged a memorable visit to the beautiful Ghent City Hall, where we received an official welcome.

I want to express my warm appreciation to Joke and Diderik in particular and to Ghent University more generally for their wonderful hospitality, to the Research Foundation – Flanders that supported the congress, and to Lucy Fleet of Springer (the successor to Reidel and Kluwer in Dordrecht) for her ongoing support of the book project. Finally, thanks, of course, to the many contributors of papers to the conference, several of which appear here in slightly revised form.

THOMAS NICKLES

UNEXPECTED DISCOVERIES, GRADED STRUCTURES, AND THE DIFFERENCE BETWEEN ACCEPTANCE AND NEGLECT

Hanne Andersen
Department of Science Studies
University of Aarhus
hanne.andersen@ivs.au.dk

In June 1934 the Italian physicist Enrico Fermi published a paper in *Nature* entitled "Possible Production of Elements of Atomic Number higher than 92" (Fermi, 1934b). In this paper Fermi reported that by bombarding uranium with neutrons he and his team had produced an element which could be element number 93, that is, a transuranic element.

Two objections followed very quickly. One objection came from von Grosse and Agruss who pointed out that different chemical properties were to be expected from element number 93 than those displayed by the element produced by Fermi (von Grosse and Agruss, 1934a, 1934b). Hence, they suggested to recategorize the element as number 91. The other objection came from Ida Noddack (1934b), who also questioned Fermi's assumptions regarding the chemical properties of element 93 and suggested that the uranium nucleus could have split into several larger fragments which would be isotopes of known, light elements.

Although Fermi had formulated his findings very cautiously,[1] it was widely accepted within the scientific community that element number 93 had actually been produced. The two objections were only partly recognized. Meitner and Hahn tested the hypothesis raised by von Grosse and Agruss that the produced element could be protactinium—and proved the hypothesis wrong (Hahn and Meinter, 1935a, 1935b)—but nobody cared for the discussion of which chem-

[1]Fermi's wording was that the results "[suggest] the possibility that the atomic number of the element may be greater than 92" and that the evidence for concluding that it be element number 93 "cannot be considered as very strong" (Fermi, 1934b, p. 899).

J. Meheus and T. Nickles (eds.), *Models of Discovery and Creativity*,
DOI 10.1007/978-90-481-3421-2_1, © Springer Science + Business Media B.V. 2009

ical properties were to be expected of element 93. Noddack's objection was simply ignored. Neither her querying the chemical properties of element number 93, nor her proposal of the division of the nucleus were discussed—or even mentioned—by other scientists working in the field.

Four years later, the hypothesis was raised once more—now by Hahn and Straßmann—that the nucleus had split into two fractions (Hahn and Straßmann, 1939a). But this time the suggestion was not ignored, on the contrary, it received an immediate, overwhelming attention and was unreservedly accepted.

Several historians of science as well as some of the historical actors have later dealt with the issue why Noddack's suggestion was ignored while Hahn and Straßmann's was accepted. Their interpretations of Noddack's proposal vary considerably. Among the historical actors looking back, Glenn Seaborg says of Noddack's paper that it "intimated the possibility of the nuclear fission reaction" (Seaborg, 1989, p. 379), while Straßmann, on the contrary, calls her suggestion a mere "accidental hit".[2]

A similar divergence of opinion can be found among the historians. Herrmann rhetorically asks if Noddack's suggestion can "be taken as the prediction of nuclear fission, as is sometimes advocated? Not really, because Ida Noddack herself does not consider her suggestion of a novel nuclear process to be meaningful enough to test it experimentally" (Herrmann, 1995, p. 53). Van Assche, on the contrary, asks "[a]s seen now, the whole publication was a recipe to discover fission, an experimental discovery that took another four years to be made and understood. How was it possible that this advice was ignored?" (van Assche, 1988, p. 206).

This confusing pattern of interpretations reflects some fundamental, recurring philosophical questions regarding unexpected discoveries, such as: Which are the constraints that make a discovery unexpected? If these constraints preclude the phenomenon, when is it rational to violate them? And is it possible that different people can rationally operate with non-identical constraints? In the following I shall give a brief account of the discovery of nuclear fission,[3] focusing on the objections to Fermi's results in 1934 and the hypothesis raised by Hahn and Straßmann in 1938/39. I shall base my account on an analysis of conceptual structures and argue that these show individual differences that may explain how different scientists can come to operate with non-identical constraints.

[2]Orig. "Zufallstreffer" (Krafft, 1981, p. 210).

[3]For an extended account of the discovery of nuclear fission, see Andersen, 1996.

1. The Conceptual Analysis

The conceptual structures of interest in this historical development are mainly taxonomic. In my analysis I shall draw on the theory of taxonomic concepts which has been developed by Kuhn. I shall argue that on the background of this theory it can be explained not only how anomalies may trigger various kinds of discoveries, but also that differences between the conceptual structures of individual scientists may explain the diverging assessments of such anomalies and on this background why some scientists accept a discovery while others reject, neglect or ignore it.

According to Kuhn's theory, a taxonomic conceptual structure is established by grouping objects into similarity classes.[4] This grouping is not determined by necessary and sufficient conditions, but by similarity between the objects within the category and difference to objects from contrasting categories. Importantly, there are no restrictions on which features can be used to judge the objects similar or dissimilar. On the contrary, anything one knows about those objects can be used in the classification. But basing a taxonomy on similarity and difference instead of explicit definitions only works if it can be assumed that no objects fall between the similarity classes. If an object does, that is, if judged by different features it seems to belong to two contrasting categories, it violates the expectations regarding which objects exist and how they behave, in short, it is an anomaly. Such anomalies may be of different sorts. They may suggest that the objects of a given category within the taxonomy behave differently than expected, but without suggesting changes to the boundaries of other categories in the taxonomy. Or they may suggest that yet another category exists within the taxonomy, but that this is simply an additional category of previously undiscovered objects such that the new category does not affect the boundary of the previously known categories. Or, most severely, they may suggest that the previously assumed category boundaries do not hold, that is, that the taxonomy must be restructured in order to work consistently. Whereas the two former kinds—changes in the characteristic features of a given category and addition of a new category to an existing taxonomy—are changes that can be assimilated within the existing taxonomic structure, the latter kind changes the taxonomic structure itself.

As it has often been pointed out, dramatic changes are only made if the triggering anomaly is somehow felt to be severe. According to the similarity account of taxonomic concepts, the severeness of an anomaly is connected to a phenomenon called *graded structures*. On a similarity account of concepts, all instances of a concept need not be equally good examples. On the contrary,

[4] This account will have to be very brief. For a full account, see e.g. Andersen et al., 1996; Chen et al., 1998; Nersessian and Andersen, 1997.

some instances may be better examples than others by being more similar to each other or more clearly dissimilar to instances of contrasting concepts. This variation in the status of instances is called a concept's 'graded structure'.[5]

These graded structures may explain why not all anomalies are equally severe. If an object is encountered that, judged from different features, is a *good* example of two contrasting concepts, this will be a severe anomaly, as it clearly questions the adequacy of the conceptual structure. On the contrary, if an object is encountered that, judged from different features, is a *poor* example of two contrasting concepts it may not call the conceptual structure in question, but just suggest that further research may be necessary to find out whether a new category exists or whether the existing categories may show some additional features that allow the objects to be unequivocally assigned to one of them. An analysis of graded structures may thus explain why a given anomaly is judged severe or unimportant, and thus why a restructuring of the taxonomy is accepted or not.

In the following I shall present an analysis of the graded structures of the concepts involved in the discovery of nuclear fission in order to explain the reactions to various anomalies and to the different claims to new discoveries.

2. Nuclear Physics

At the beginning of the 1930s the nucleus was conceived of as a collection of individually existing protons, electrons and α-particles (Gamow, 1931). After the neutron was discovered in 1932, the nuclear electron hypothesis was no longer necessary, and the nucleus was conceived as existing of protons and neutrons which possibly clustered together in α-particles.[6]

In accordance with the view of particles existing individually within the nucleus, Gamow had developed in 1928/29 a quantum mechanical theory of α-decay in which he treated nuclear disintegration as a tunnelling phenomenon (Gamow, 1929a, 1929b). On this theory, only particles up to the size of the α-particle were energetically capable of tunnelling the potential barrier.

In 1934 Curie and Joliot discovered that they could induce radioactivity in light elements by bombarding them with α-particles (Curie and Joliot, 1934). Due to the potential barrier, α-particles could only be used for bombarding light elements, and Fermi therefore suggested to use the electrically neutral neutron as projectile instead. Fermi and his collaborators started with a systematic investigation, "irradiating all the substances [they] could lay [their] hands on" (Segré, 1970, p. 75). They reported that for a large number of elements of any

[5]See e.g. Barsalou, 1992, ch. 7.3.2 and Lakoff, 1987, ch. 2 for an overview of the psychological literature on graded structures.

[6]For an account of the nuclear electron hypothesis, see Stuewer, 1983.

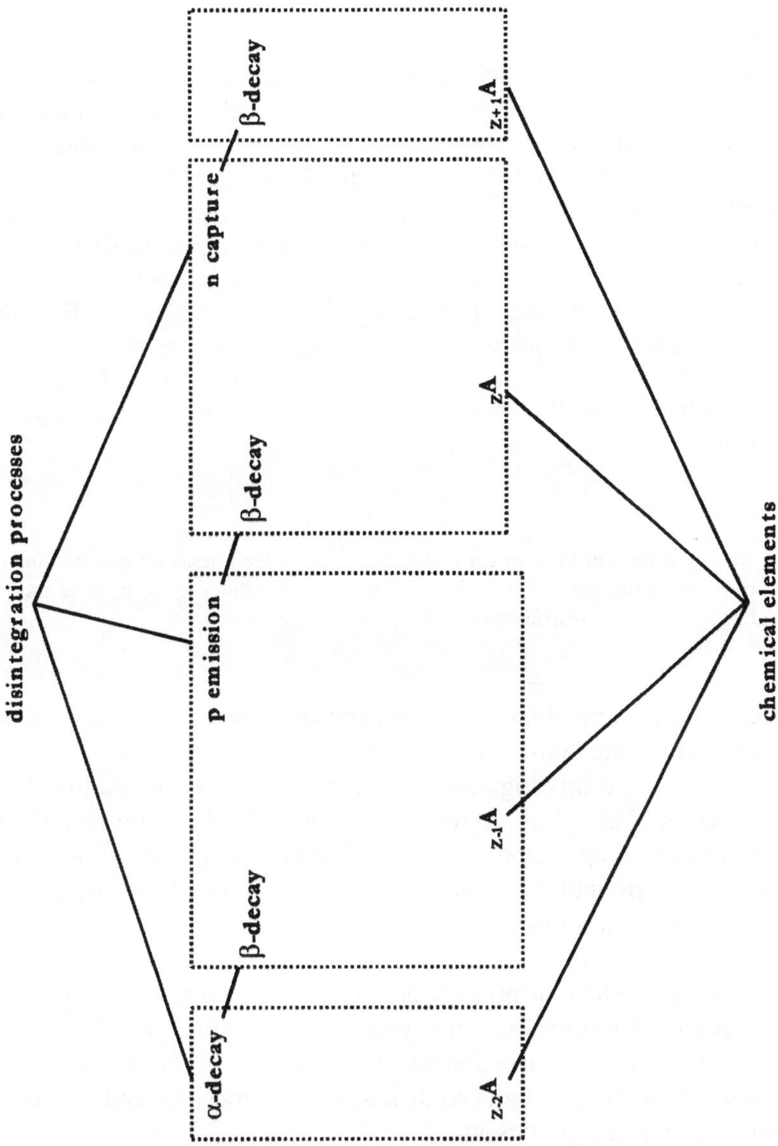

Figure 1. Main taxonomy of disintegration processes and its connection to the taxonomy of elements.

Wir geben nun eine Übersicht der erzwungenen Umwandlungen. Bei jedem Reaktionstyp sind nur die Eigenschaften aufgeführt, die ihn von den übrigen Prozessen unterscheiden; soweit keine besondere Angabe gemacht ist, gelten also die obengenannten allgemeinen Regeln.

(α, d). Scheint nur in einem einzigen Fall beobachtet zu sein (75). Daß das Deuteron als unzerlegter Kernsplitter auftritt, ist wegen seiner geringen Stabilität in der Tat unwahrscheinlich.

(α, p). An diesem Prozeß wurde die künstliche Kernumwandlung durch Rutherford und Chadwick entdeckt. Er ist energetisch besonders günstig, wenn der Ausgangskern ungerades Z und gerades N hat, da er zur Erhöhung von Z um eine, von N um zwei Einheiten führt.

(α, n). Bei diesem Prozeß wurde das Neutron entdeckt. Er ist energetisch günstig bei Kernen mit geradem Z und ungeradem N.

(α, γ). Nicht beobachtet. Da der Prozeß an sich mit großer Energieausbeute möglich sein sollte, muß die Erklärung in der geringen Emissionswahrscheinlichkeit der γ-Strahlung liegen (§ 34). Einfangung ohne Emission einer Strahlung ist (§ 4) nach den Erhaltungssätzen unmöglich.

Figure 2. Extract from von Weizsäcker's *Die Atomkerne* which treats all possible induced radioactive processes in the form of a list of all possible permutation of p, n, d, α and γ as projectile and decay products, respectively.

atomic weight, neutron bombardment would produce unstable elements which disintegrated through the emission of β-particles (Fermi, 1934b).

The next step was to investigate the primary processes that lead to the β-radiating elements. The original group consisted of Fermi, who had already achieved international reputation as a theoretical physicist, and the two physicists Amaldi and Segré, but they soon recruited the chemist D'Agostino in order to make the chemical separations necessary for identifying which elements were produced in the disintegration processes. Identification of the produced elements would then reveal the primary process by which it had been produced. The group reported that three main processes were possible: α emission, proton emission and neutron capture (Fermi, 1934b, p. 898).[7] This established the main taxonomy of artificially induced disintegration processes and its connection to the taxonomy of elements (fig. 1).

[7] α emission was identified for Al, Cl and Co, proton emission for Ph, S and Zn, and neutron capture for Br and I.

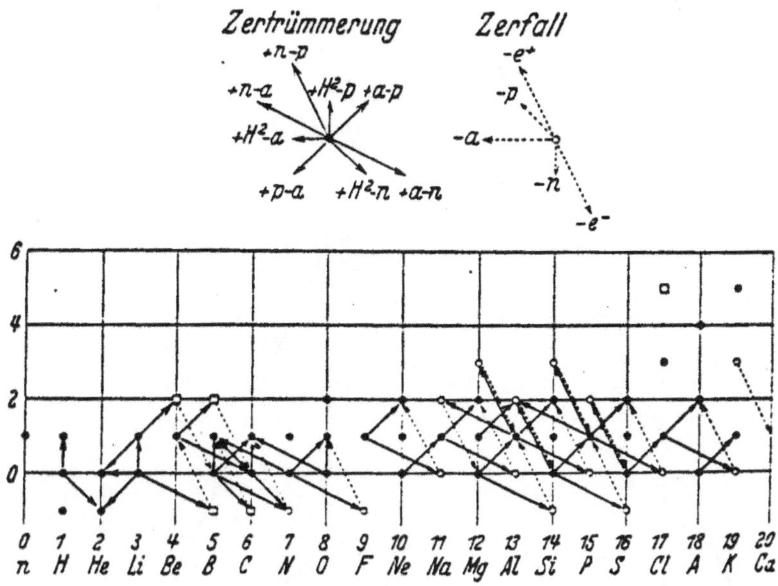

Figure 3. Checker-board like diagram of possible nuclear transmutations. From Meitner, 1934.

This taxonomy was in fine accordance with Gamow's theory of decay which precluded decay products larger than the α-particle.[8] In the years that followed, Gamow's result that only particles up to the size of the α-particle could be emitted would become tacitly accepted in the whole scientific community to such an extend that the mere possibility of larger decay products would never be mentioned (fig. 2).

Likewise, the diagrams and notations which were developed could only represent the idea that a projectile hits a nucleus which as a result transformed into another nucleus by the emission of a particle (fig. 3). The range of the taxonomy seemed well-defined.

Having established this taxonomy of artificially induced disintegration processes, Fermi and his team took special interest in heavy nuclei. The general instability of the heaviest elements might give rise to successive β-decays, and possibly that could lead to a transuranic element (fig. 4).

When they bombarded U with neutrons they discovered at least 5 different disintegration processes with different half-lives: 10 sec., 40 sec., 13 min. plus at least two more periods from 40 minutes to one day (Fermi, 1934b, p. 899). But where did they belong (fig. 5)?

[8] Fermi referred explicitly to Gamow's work in several papers, see e.g. Fermi et al., 1934, 1935.

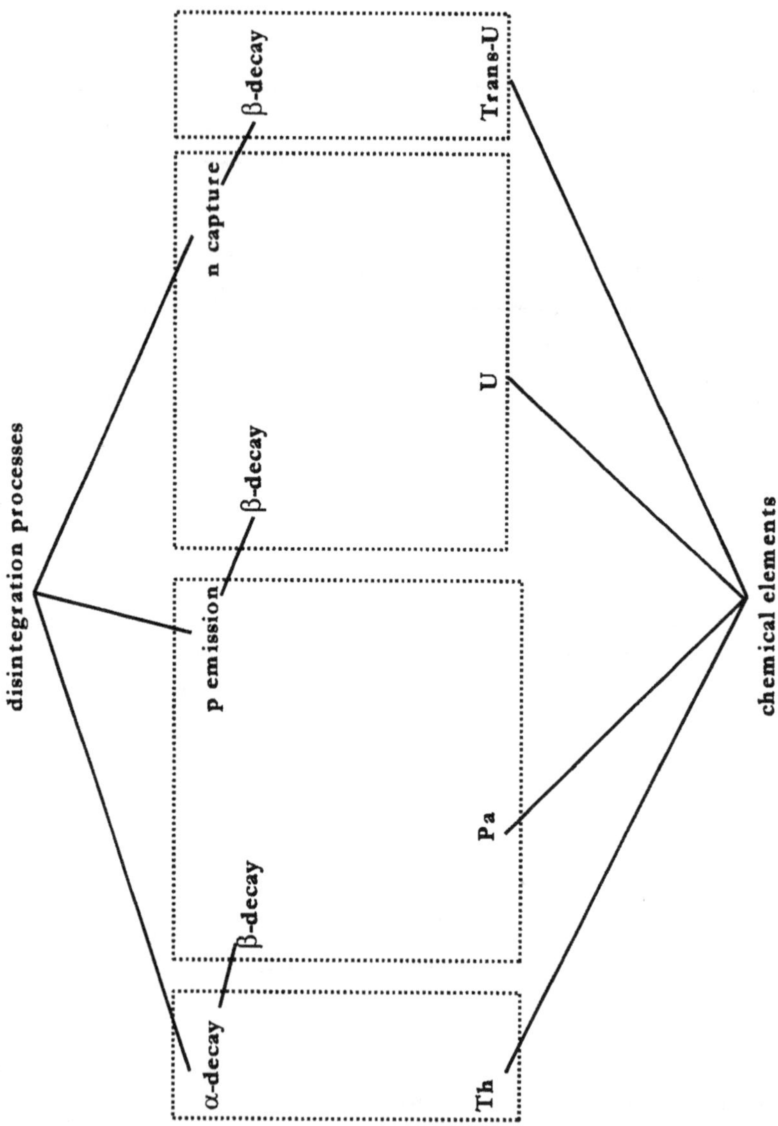

Figure 4. Taxonomy of disintegration processes when bombarding uranium with neutrons and the connection to the taxonomy of heavy elements. The β-decay series following the primary disintegration processes might give rise to transuranic elements.

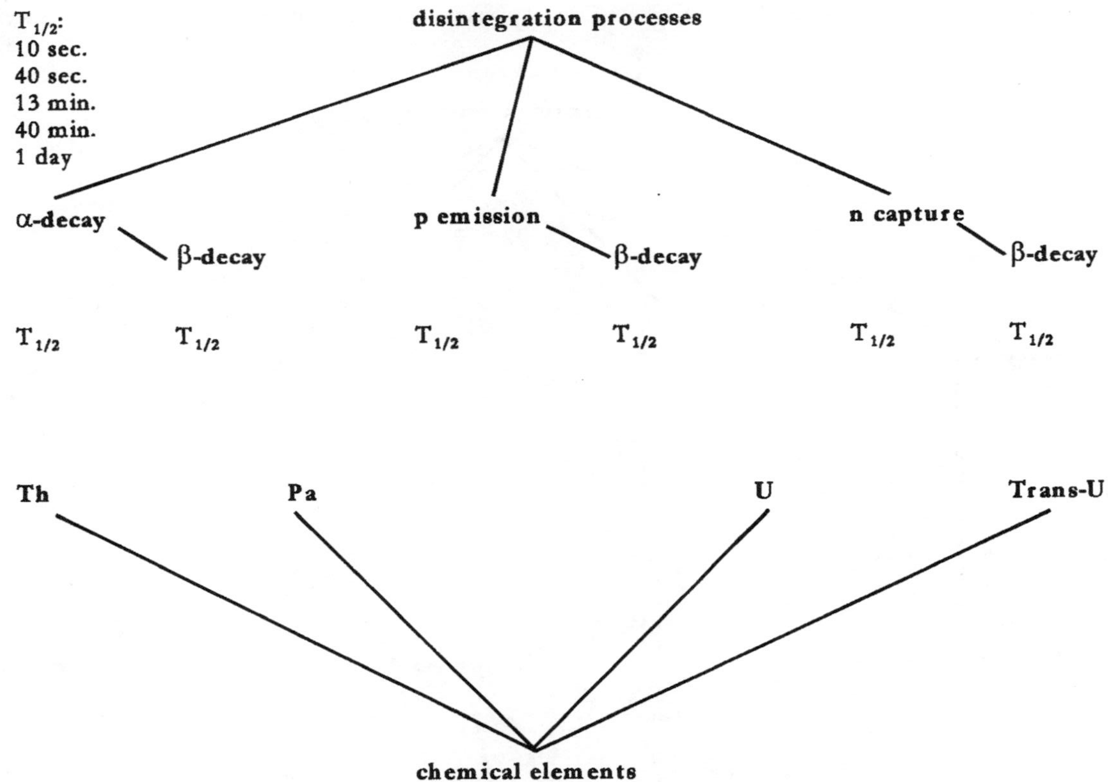

Figure 5. Half-lives could be used to identify different disintegration processes, but first it had to be determined which disintegration processes had which half-lives.

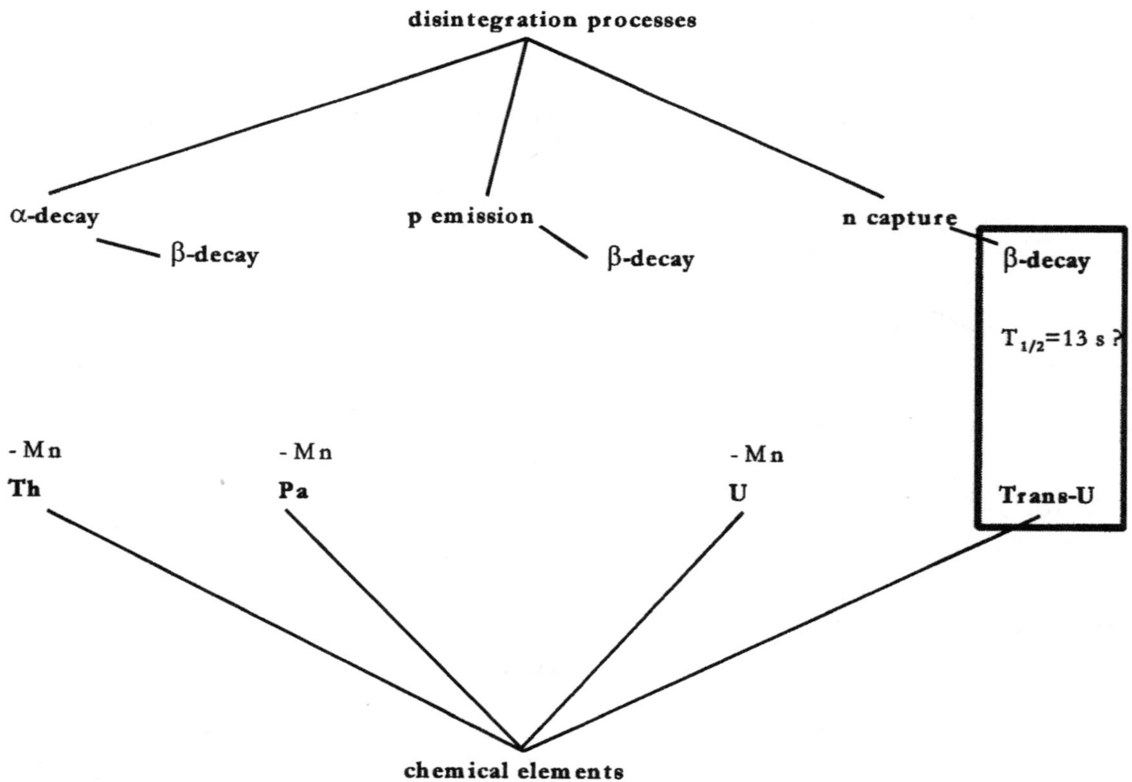

Figure 6. Manganese precipitation processes showed that the element produced in the 13 min. process could not be any of the known heavy elements (indicated in the diagram as -Mn). It was therefore categorized as a transuranic element.

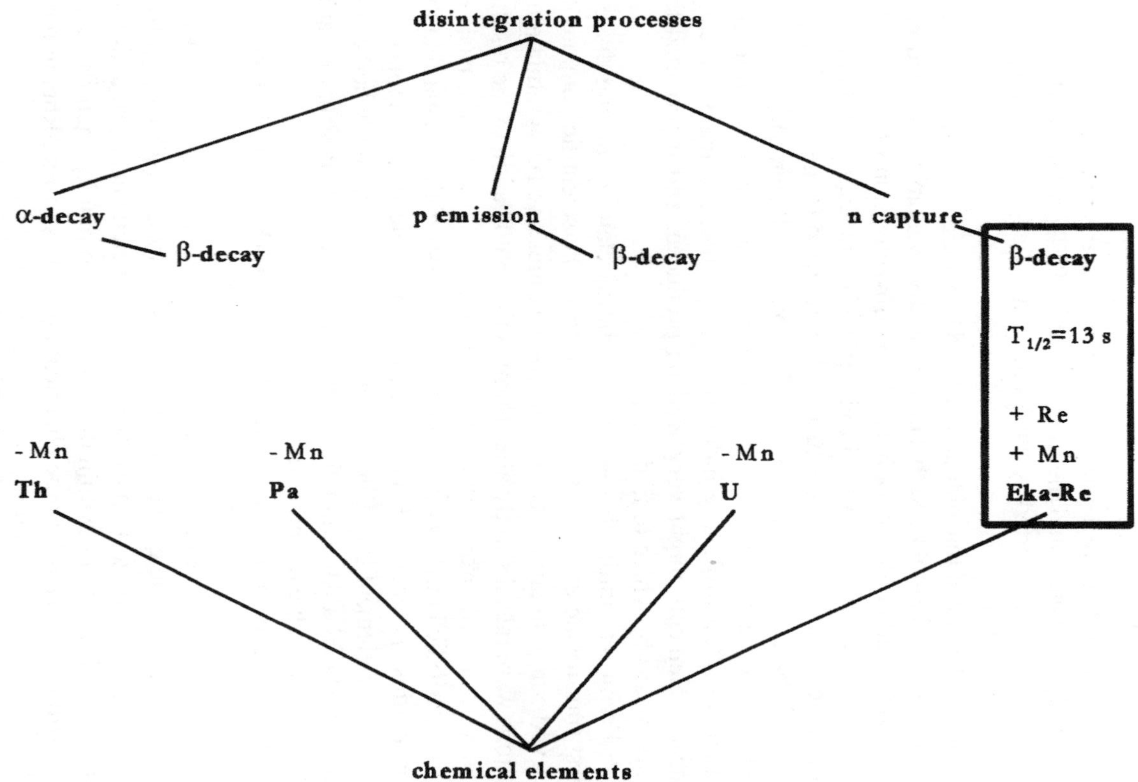

Figure 7. Since the element produced in the 13 min. was chemically homologous with manganese and rhenium (indicated in the diagram as +Mn and +Re) it was hypothesized to be element number 93 and belong to the same subgroup in the periodic table as rhenium. It was given the name Eka-Re.

Concentrating on the element with the period of 13 min., they showed that a manganese precipitation process would separate this element from "most of the heaviest elements" (Fermi, 1934b, p. 899), and they concluded that "this negative evidence about the identity of the 13 min. activity from a large number of heavy elements suggests the possibility that the atomic number of the element may be greater than 92" (fig. 6).

They hypothesized that "if it were an element 93, it would be chemically homologous with manganese and rhenium" (Fermi, 1934b, p. 899) and reported that this hypothesis was supported by the results of another precipitation process using rhenium sulphide (fig. 7).

However, they also noted that elements 94 and 95 would probably not be easy to distinguish from element 93 and that consequently "valuable information on the processes involved could be gathered by an investigation of the possible emission of heavy particles" (Fermi, 1934b, p. 899). Hence, given that chemical characteristics might not be conclusive, they referred to the desirability of including decay characteristics in the classification as well (fig. 8).

The discovery of transuranic elements was therefore a very *expected* discovery. The taxonomy of artificially induced disintegration processes indicated that transuranic elements might very well be produced, and it provided the classificatory means by which to find them.

Although Fermi was initially very cautious in his claim of having discovered the first transuranic element, the reaction from the scientific community was unreserved congratulations. Or, rather, *almost* unreserved congratulations. Two objections were raised shortly after Fermi's first publication of the results. The first came from von Grosse and Agruss (1934a, 1934b). On the basis of Mendelejeff's periodic law they questioned Fermi's assumptions regarding the chemical properties of element 93. However, what they questioned in this paper was solely how the element Eka-Rhenium would behave,[9] but not whether the element 93 would be Eka-Rhenium, that is, whether it would belong to the same group in the periodic table as rhenium (fig. 9).

Von Grosse and Agruss further criticized the process which Fermi's team had used to rule out protactinium,[10] and reported that according to their experiments the new element could very well be protactinium. However, Meitner and Hahn showed that this was not the case (Hahn and Meinter, 1935a, 1935b) (fig. 10).

Whereas the specific suggestion to recategorize the element as protactinium was discussed—and rejected—within the scientific community, their criticism of Fermi's assumptions regarding the chemical properties of Eka-Rhenium re-

---

[9]More specifically, von Grosse and Agruss questioned whether the highest oxide of Eka-Rhenium would form an acid under the conditions of Fermi's experiment, or whether it would precipitate with the manganese carrier.

[10]Fermi had used a very short-lived isotope of protactinium which made the chemical operations very difficult.

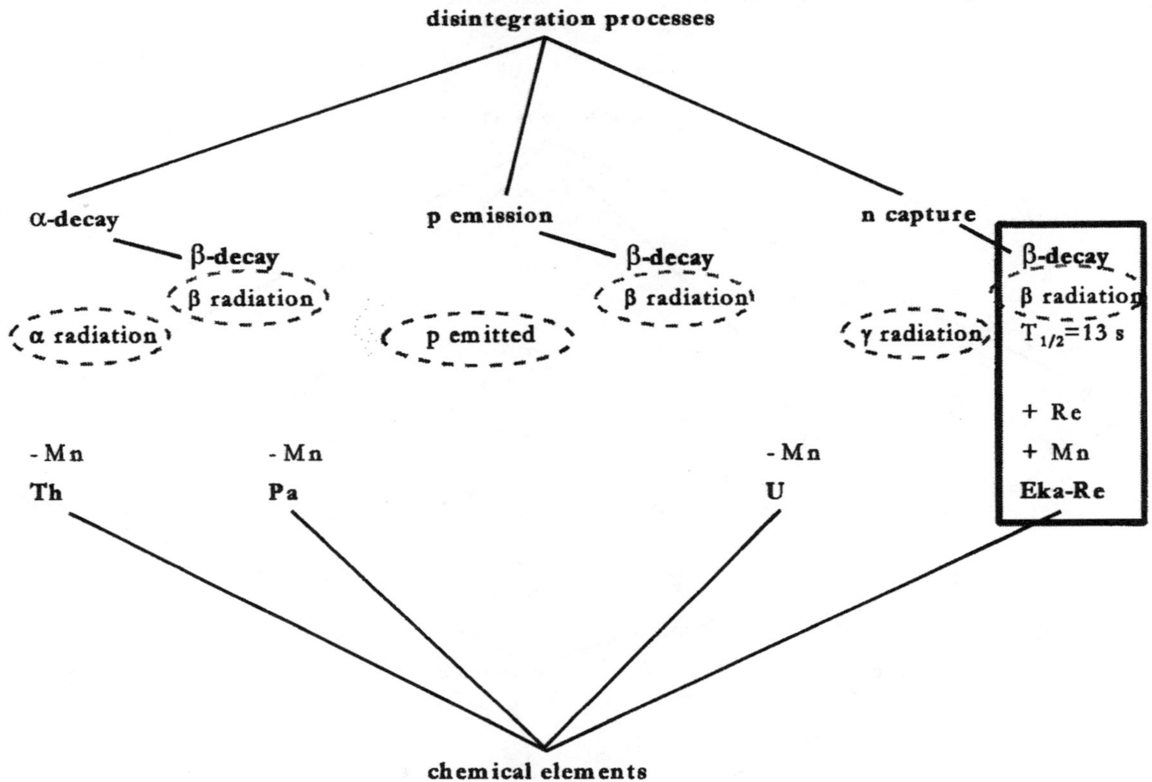

Figure 8. Since it was difficult to categorize the transuranic elements only on the basis of chemical behaviour, including other characteristics from the taxonomy of disintegration processes, such as radiation, would be advantageous.

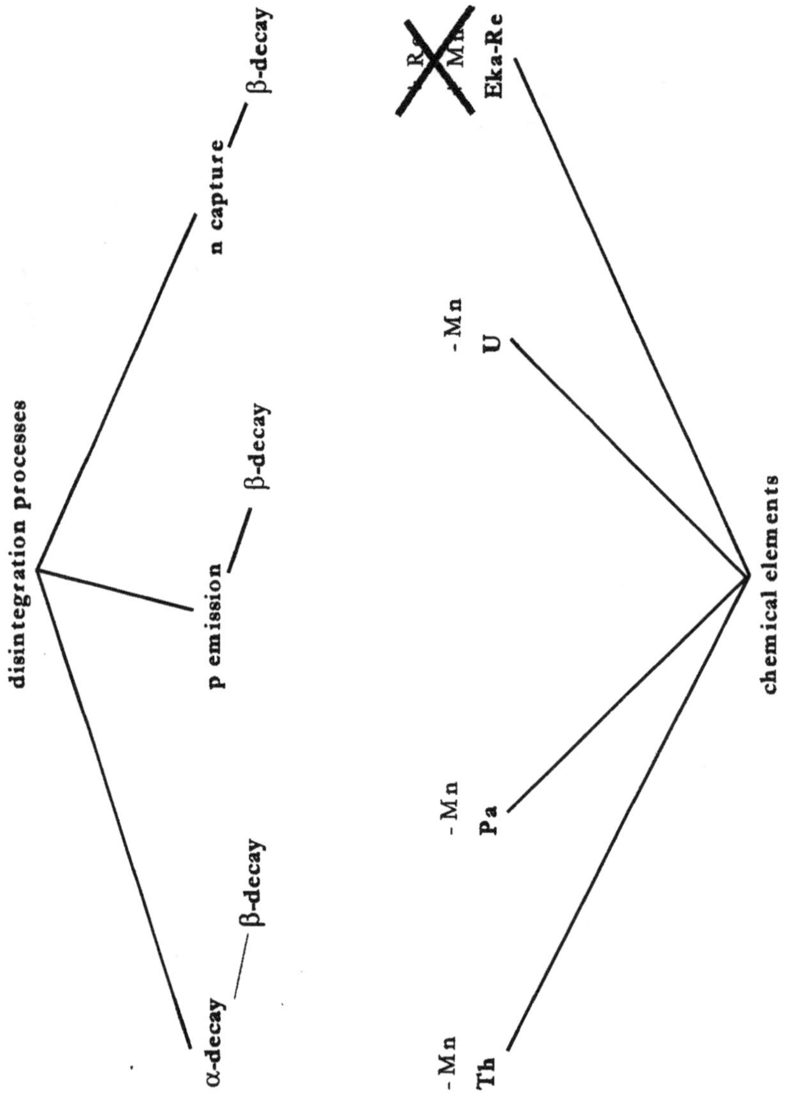

Figure 9. Von Grosse and Agruss questioned whether element 93 would be chemically homologue to rhenium and manganese.

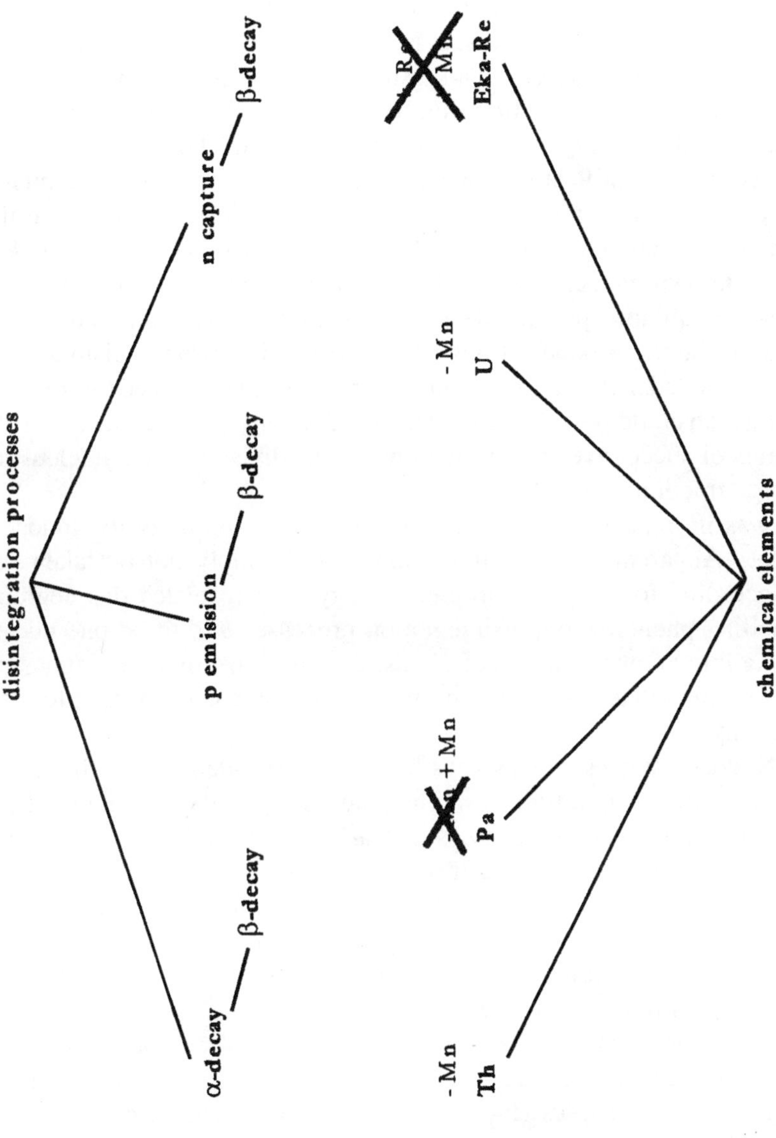

Figure 10. Von Grosse and Agruss reported that according to the precipitation processes they had conducted, the new element could be protactinium.

mained unnoticed. Von Grosse published a paper a few months later in which he both substantiated his claim regarding Eka-Rhenium, and also pointed out that element 93 might not even be Eka-Rhenium, but could instead belong to a transition group which would imply a completely different set of chemical properties (von Grosse, 1934) (fig. 11).[11]

Still, there were no indications in any of the papers from the Rome or the Berlin teams that they seriously discussed whether element 93 would have the chemical properties which Fermi had assumed in his classification.

The second objection came from Ida Noddack (1934b). She too pointed out that element number 93 might not have the chemical characteristics which Fermi had assumed in his identification, especially regarding the rhenium precipitation process. However, the alternative she suggested was much more radical than the alternative which von Grosse and Agruss had proposed. She pointed out that several known elements would behave like Fermi's new element in the manganese precipitation process. But these were all much lighter than uranium and could not be the product of any of the artificially induced disintegration processes contained in Fermi's taxonomy. She therefore suggested two different processes which could possibly lead to the production of light elements: either a long series of successive transformations, or the division of the nucleus into several large fractions (fig. 12).

There was no reaction at all from the scientific community to Noddack's suggestion. Apparently, these suggestions could simply not be taken seriously. According to Gamow's droplet analogy, which treated disintegration as a tunnelling phenomenon, disintegration processes *had* to be one nucleus transmuting into another nucleus of almost the same size by releasing a small particle. On this model, there was no way a nucleus could divide into a few large fractions.

What Noddack suggested was not filling out a well-defined gap in the taxonomy like Fermi's suggestion was. The potentiality of the taxonomy of artificially induced disintegration processes clearly did not include the division of the nucleus. Whereas discovering transuranic elements was highly expected, discovering that the nucleus had split into large fractions would not only be unexpected, it would be highly revolutionary, demanding changes in the principles underlying the taxonomy. This did not seem necessary, neither to Fermi and his group, nor to anybody else.

During the four years to follow, several discoveries were made that had not been expected, but which could all be included in the taxonomy without changing its underlying principles (fig. 13). The process 'neutron chipping' was introduced as a simple addition to the taxonomy which had not been expected

[11] von Grosse assumed the transition group to start with protactinium, hence the transuranic elements would have chemical properties similar to this element.

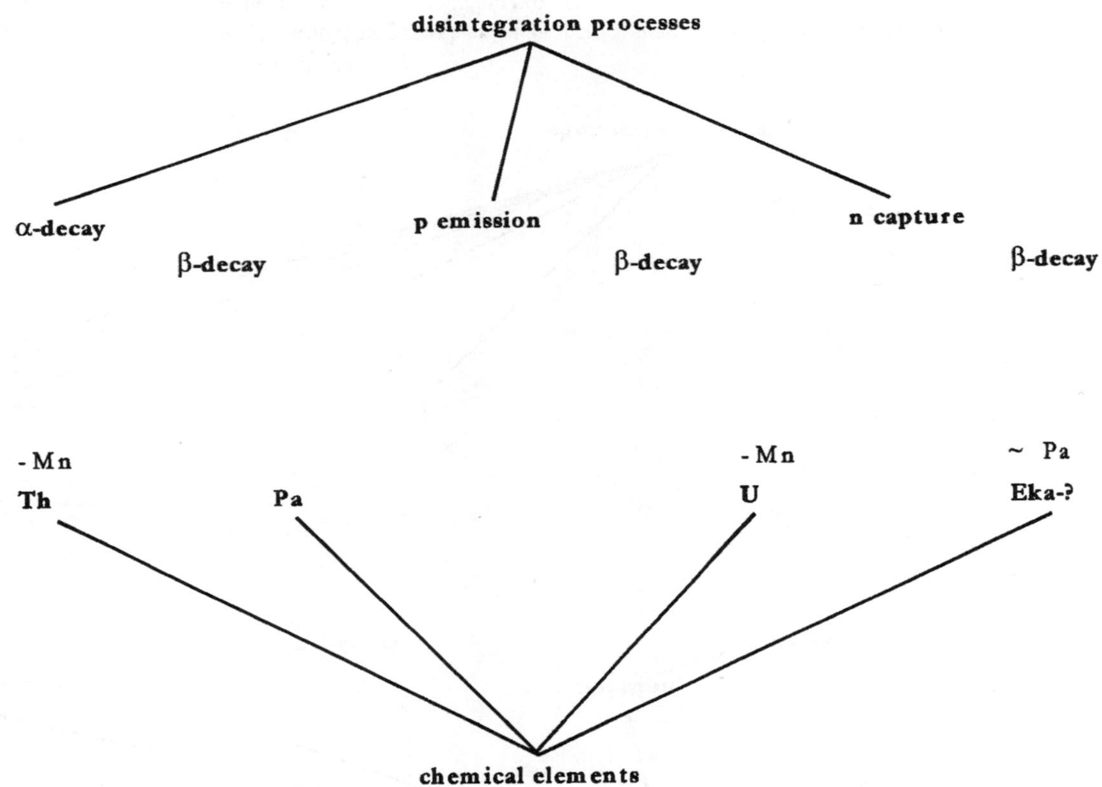

Figure 11. In a later paper von Grosse argued that element number 93 could belong to a transition group like the lanthanides. This could mean that element number 93 had chemical characteristics similar to those of protactinium.

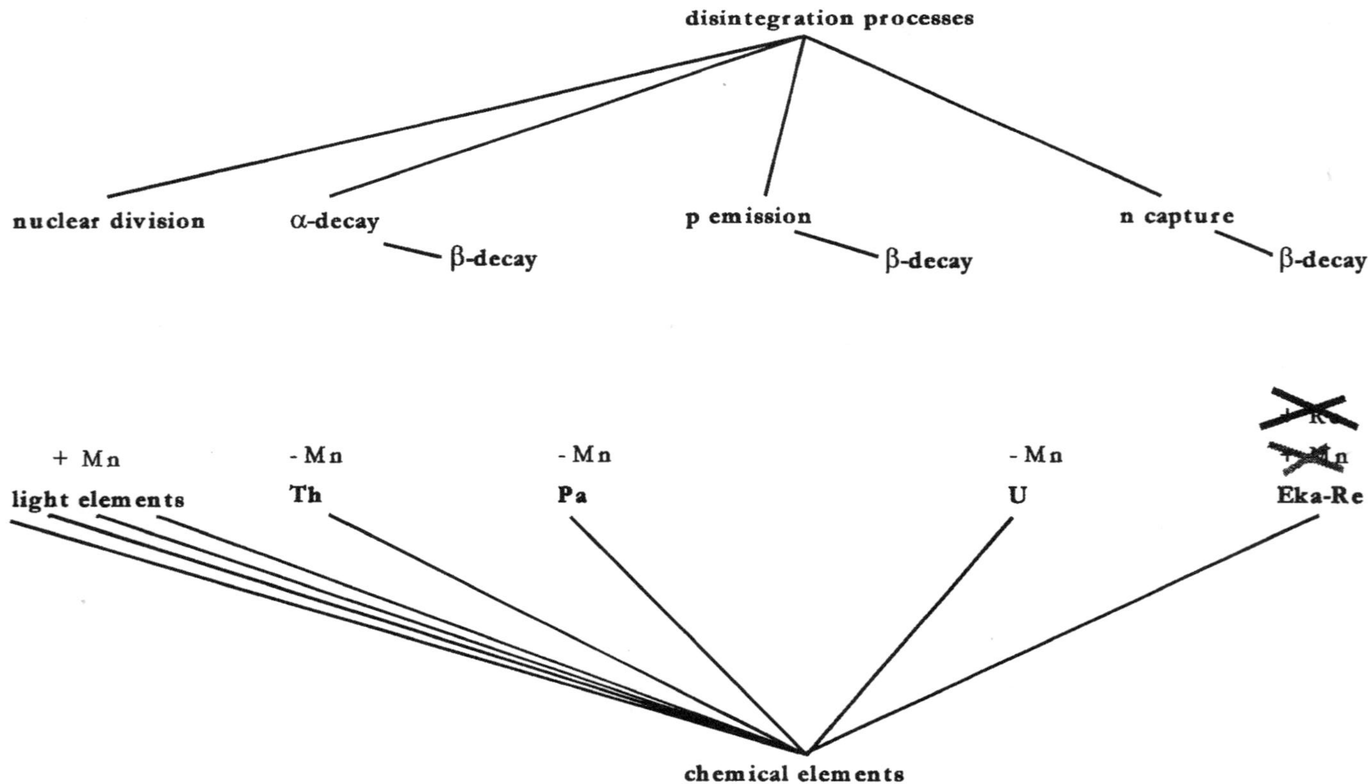

Figure 12. Noddack questioned if element 93 would have the chemical properties suggested by Fermi. She also pointed out that several light elements would behave like Fermi's new element in the manganese precipitation process.

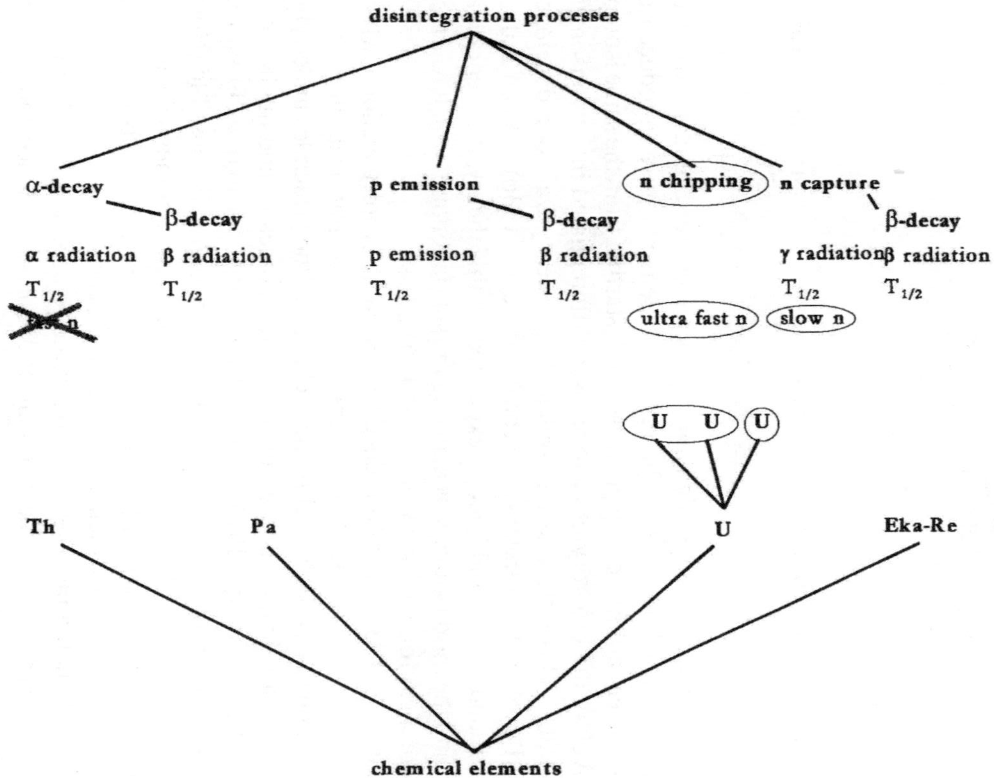

Figure 13. Several changes were made to the taxonomy of disintegration processes and the connection to the taxonomy of chemical elements during the period 1934-38. Neutron chipping and isomers were introduced as new concepts, and various features, such as the energy of the projectile neutron, were included in the classification.

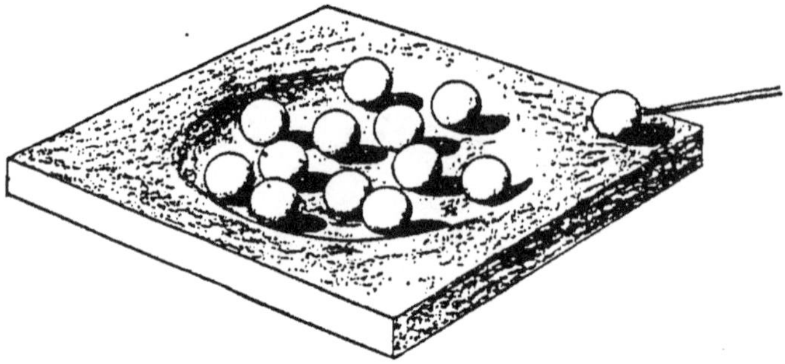

Figure 14. Illustration from Bohr, 1937, depicting the nucleus as a basin with a number of balls in it.

but which could be made unproblematically (Meitner and Hahn, 1936). Various features by which to classify the produced elements were introduced or abandoned. When multiple decay processes both starting from the same isotope were discovered, the category of isotopes was differentiated into two isomers (Meitner, 1936). This started as a postulate, but a theory was soon developed which could explain the subdivision (von Weizsäcker, 1936). When a third isomer was found which could not be explained by this theory, the theory was abandoned and the third isomer was added to the sub-taxonomy which was then without theoretical foundation (Meitner et al., 1937).

Importantly, in 1936 Bohr introduced a different droplet analogy which treated the nucleus like an oscillating droplet and disintegration as an evaporation phenomenon (Bohr, 1936). This was a change of the principles underlying the taxonomy of artificially induced disintegration processes and might therefore change the taxonomy. Bohr explicitly noted that for extremely violent impacts, an explosion of the whole nucleus was theoretically possible (Bohr, 1936, p. 348). However, although this new droplet analogy thus widened the potentialities of the taxonomy, the explosion of the nucleus was only introduced as a potential phenomenon which did not need to exist.

Bohr explicitly noted that the energy required to achieve an explosion was "far beyond the reach of experiment" (Bohr, 1936, p. 348). Hence, there would not be much reason in looking for the new phenomenon. Further, although Bohr did not specify what the end product of an explosion would be, judging from the illustrations he used, depicting the nucleus as a basin with a number of billiard balls in it (fig. 14), it seems likely that he thought of the explosion fragments as small particles, not large fractions.

Hence, even if the energy required to achieve an explosion became within reach, the phenomenon which Bohr expected was not that much different from

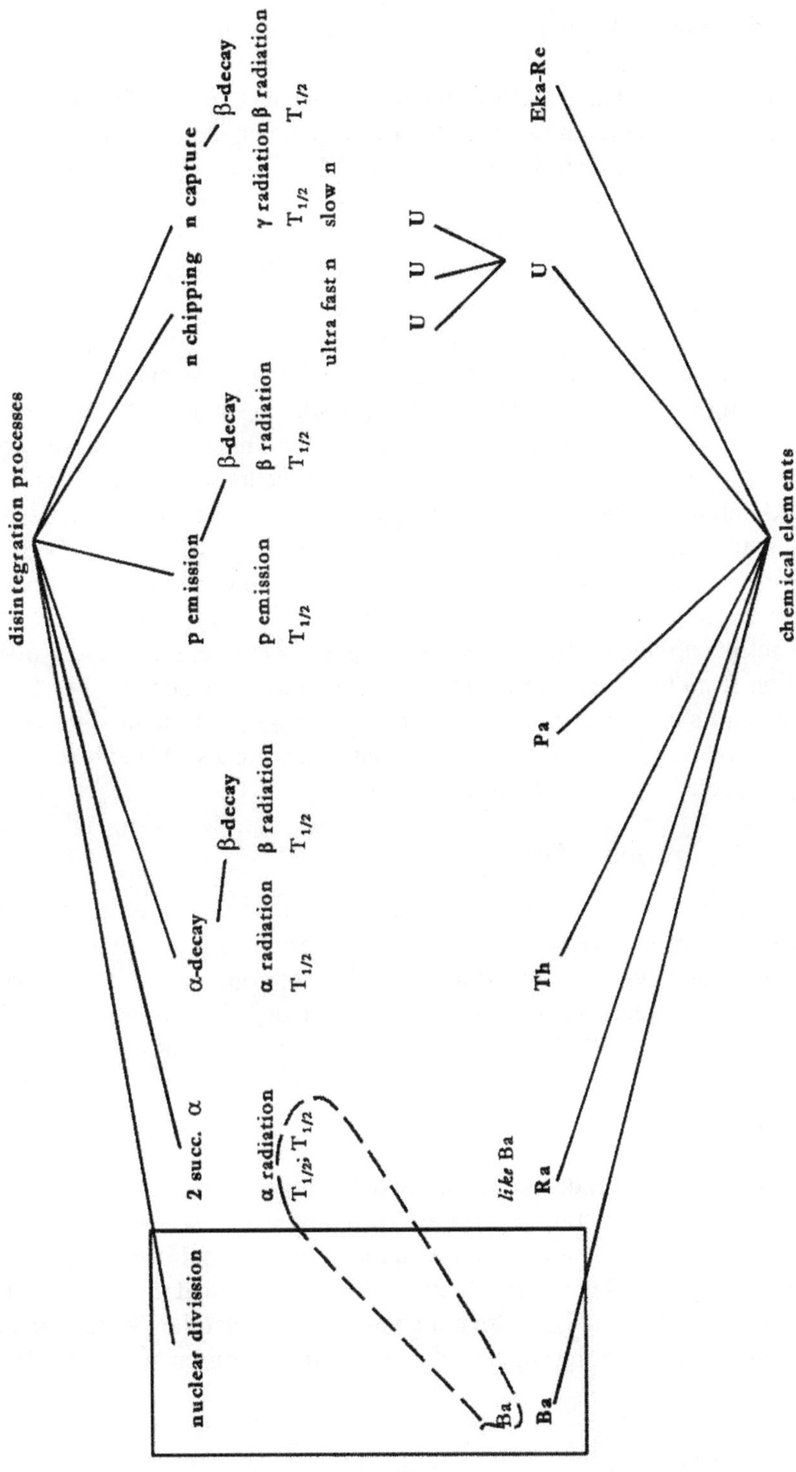

Figure 15. Hahn and Straßmann discovered that what they had categorized as radium had the chemical characteristics of barium.

the processes already known. In effect, nobody took notice of the fact that Bohr's model could widen the potentiality of the taxonomy.

By the end of 1938, Hahn and Straßmann discovered that one of the produced daughter elements could not be separated from the much lighter carrier element they had used in the precipitation process. The taxonomy of the elements did not fit with the taxonomy of artificially induced disintegration processes, but contrary to the anomalies pointed out by Noddack and by von Grosse and Agruss, this time the anomaly involved a very well-investigated part of the taxonomy of the elements (fig. 15).

Hahn resorted to the same shift as Noddack had done: postulating a new disintegration process. He wrote to Meitner, suggesting that the nucleus had been divided, asking if she could figure out an explanation (Hahn, 1975, pp. 77ff.).

Meitner discussed it with her nephew Frisch, who had worked on the consequences of Bohr's droplet analogy.[12] They immediately realized that this model did carry the potential for the phenomenon of nuclear fission, which could be explained by violent oscillations of the droplet (Meitner and Frisch, 1939). Hence, although the discovery of nuclear fission was not predicted or in any way expected, it turned out to fill a gap in the taxonomy whose existence just had not been realized before. Since the gap was there, ready to be filled, the reaction from the scientific community was immediate acceptance. Some even expressed a certain annoyance that this potentiality of the taxonomy had not been realized before. "What idiots we have been that we haven't seen that before" was Bohr's reaction (Weart, 1983, p. 113).

3. Philosophical Morals

What does a historical case like this one tell us? On the face of it, it may seem as if the difference between Noddack and Hahn/Staßmann was that Noddack did not really know what she was doing and therefore did not realize that what she suggested was—according to contemporary theory—impossible. As Hahn and Straßmann put it in 1939, the question of whether the supposed transuranic elements could instead be fragments of an exploded nucleus "could not be posed before the totally unexpected fission process as it was inconsistent with the general conceptions of nuclear physics of that time" (Hahn and Straßmann, 1939b, p. 451). However, this is definitely too simple.

In a way the question could have been meaningfully posed since 1936, but nobody had linked Bohr's new model with the radiochemical investigations on transuranic elements. Hence, when they wrote to Meitner that it looked like the nucleus had split, Hahn and Straßmann knew as little what they were doing as Noddack had done when she suggested that the nucleus might have split. Both

[12]See e.g. Stuewer, 1994.

the chemist Noddack and the chemists Hahn and Straßmann had seen a chemical anomaly and suggested to resolve it by changing the taxonomy of disintegration processes, without any of them knowing how the changed taxonomy and the underlying theory could be brought to comply with each other again.

Hence, the difference cannot be described as Noddack violating a constraint she was ignorant of while Hahn and Straßmann were not. However, there does seem to be some sort of difference regarding the constraints they disregarded (whether deliberately or due to ignorance). Whereas in Noddack's case there was no theory available to support her suggestion by explaining the new taxonomy, in Hahn's and Straßmann's case there was. However, this is only part of the explanation. Changes in taxonomy may be accepted although there is no theory available to explain them. One example is the third isomer form which could not be explained by the theory, but was still included in the taxonomy. Another example are some very long β-decay series which were also accepted although they were deeply at variance with theory. Whereas the availability of an explanation explains the *immediate* acceptance of Hahn and Straßmann's discovery, the neglect of Noddack's suggestion cannot be explained by the lack of an explaining theory alone.

Instead, the severeness of the anomaly leading to the suggestion must be involved as well. Accepting a new and revolutionary discovery is not just a question of whether there is a theory available which can explain the discovery—for revolutionary discoveries usually there is not—but also a question of which problems would be solved by accepting the revolutionary discovery. Apparently, whereas Noddack saw an anomaly so severe that she suggested to change the taxonomic structure, others saw no anomaly at all.

This brings us back to the graded structures and how they may differ for different scientists. As explained in the beginning, graded structures can arise because categories are constituted by similarity relations. Since there are no restrictions on the features that may be used to judge similarity and dissimilarity, different scientists belonging to the same scientific community may use different features in identifying objects of the same categories. Different scientists emphasizing different features when categorizing the same objects may therefore develop different graded structures. As the severeness of an anomaly may depend on graded structures, it follows that different members of the scientific community who have developed different graded structures may not necessarily agree on which anomalies are severe and which are not.

Noddack was an analytical chemist who had worked for years searching for the still missing elements in the periodic table, and earlier in 1934 she had expressed her firm belief that accurate predictions of the characteristics of the transuranic elements had to come before their discovery (Noddack, 1934a, p. 304). Elsewhere she had described constraints on chemistry derived from theoretical physics as 'dogmas' that would one day be refuted.[13] To Noddack the chemical identifications clearly had much more weight in identifying the transmuted nuclei than physical expectations of possible decay series, and if chemical characteristics suggested that a new disintegration process had to be added, then so it be. It would solve some problems,[14] only at the cost of giving up a mere presupposition about what might or might not exist in an area of research which had not been entered before.

Fermi's team, on the contrary, used the conceptual scheme of disintegration processes to narrow the range of possible elements, and then only made chemical analysis within this narrow range of possibilities. To them, as well as others in the field, Noddack had not pointed to any serious anomalies with her vague chemical speculations, but only to "a lack of rigour in the argument" (Amaldi, 1984, p. 277). Von Grosse and Agruss had pointed both to an anomaly and to an alternative interpretation of the results which would dissolve the anomaly, but after their alternative suggestion was rejected, the chemical anomaly that had triggered it seemed forgotten. Nobody ever discussed whether the assumed chemical properties of element 93 were correct. Hence, no anomaly was seen, and without a serious anomaly, there was no reason to accept a radical change of a highly successful taxonomy.

Thus emphasizing very different features in their classifications, although they may previously have agreed on the categorization of all other elements, the new element was categorized differently according to the features the different scientists employed in their categorization. Here the latent differences between their criteria for categorization of elements suddenly come to notice.

Noddack emphasized chemical analysis, and when she saw a potential conflict with some physical criteria she considered unfounded, those criteria would be the ones to go. However, her criteria for the chemical analysis were rather vague, and this made it impossible for her to convey it to others, just as the fact that she did not pursue it herself may indicate that she assessed it as an anomaly that could just await its solution in due time, but not a pressing one that needed her immediate attention. Von Grosse and Agruss emphasized chemical analysis too, and they had much more elaborate expectations than Noddack. This made them pursue the anomaly, but since their interest in computing chemical properties from Mendelejeff's periodic law was not shared by the rest of the

[13]This refers to a commentary she made to the claim that isotopes of the same element are chemically identical because they have identical electronic configurations: "this claim was a dogma, however well motivated by atomic theory. It will therefore suffer from the same fate as all dogmas, that is, it will one day be refuted" (Noddack, 1934a, p. 305).

[14]Noddack had previously expressed the opinion that elements with even atomic number—94 and 96—were the most likely transuranic elements to find and that the transuranic elements would have shorter and shorter half-lives (Noddack, 1934a). Hence, both the long half-lives and the odd atomic number may have surprised her.

scientific community, they remained alone in their pursuit. Most others emphasized the physical characteristics. The chemical characteristics they used were based only on the subgroups of the periodic table, with no considerations of possible differences between the periods. As this accidentally made physical and chemical categorization fit, they saw no reason to revise the chemical characteristics; they were strongly supported by the physical characteristics. Hahn and Straßmann had also worked that way, emphasizing physical characteristics, but when an anomaly appeared in the classification of well-known elements, it was a conflict between two sets of equally important features. This was an anomaly so severe that they were ready to give up what they had previously believed in. Since most others shared their emphasis, both on physical characteristics and on the characteristics of well-known elements, the severeness of the anomaly was easily conveyed to the rest of the scientific community.

What we see here is that the similarity account of concepts presented here may both explain the unequivocal use of concepts in consensus situations, and also provide the necessary resources to explain the emergence of dissensus. This may be the way to proceed if we want to answer the questions how constraints are violated and how different people can rationally operate with non-identical constraints.

References

Amaldi, E. (1984). From the discovery of the neutron to the discovery of nuclear fission. *Physics Reports*, 111:1–332.

Amaldi, E., D'Agostino, O., Fermi, E., Pontecorvo, B., Rasetti, F., and Segré, E. (1935). Artificial radioactivity produced by neutron bombardment. *Proc. Roy. Soc.*, 149A:522–558.

Andersen, H. (1996). Categorization, anomalies, and the discovery of nuclear fission. *Stud. Hist. Phil. Mod. Phys.*, 27:463–492.

Andersen, H., Barker, P., and Chen, X. (1996). Kuhn's mature philosophy of science and cognitive psychology. *Philosophical Psychology*, 9:347–363.

Barsalou, L. (1992). *Cognitive Psychology. An Overview for Cognitive Scientists*. Lawrence Erlbaum Associates, Hillsdale, NJ.

Bohr, N. (1936). Neutron capture and nuclear constitution. *Nature*, 137:344–348.

Bohr, N. (1937). Transmutations of atomic nuclei. *Science*, 86:161–165.

Chen, X., Andersen, H., and Barker, P. (1998). Kuhn's theory of scientific revolution and cognitive psychology. *Philosophical Psychology*, 11:5–28.

Curie, I. and Joliot, F. (1934). Un nouveau type de radioactivité. *Compt. Rend.*, 198:254–256.

Fermi, E. (1934a). Artificial radioactivity produced by neutron bombardment. *Nature*, 134:668.

Fermi, E. (1934b). Possible production of elements of atomic number higher than 92. *Nature*, 133:898–899.

Fermi, E. (1934c). Radioactivity induced by neutron bombardment. *Nature*, 133:757.

Fermi, E., Amaldi, E., D'Agostino, O., Rasetti, F., and Segré, E. (1934). Artificial radioactivity produced by neutron bombardment. *Proc. Roy. Soc.*, 146A:483–500.

Gamow, G. (1929a). Über die Struktur des Atomkerns. *Physik. Z.*, 30:717–720.

Gamow, G. (1929b). Zur Quantentheorie der Atomzertrümmerung. *Z. Phys*, 52:510–515.

Gamow, G. (1931). *Constitution of Atomic Nuclei and Radioactivity*. Clarendon, Oxford.

Hahn, D., editor (1975). *Otto Hahn. Erlebnisse und Erkenntnisse*. Econ. Verlag, Düsseldorf.

Hahn, O. and Meinter, L. (1935a). Über die künstliche Umwandlung des Urans durch Neutronen. *Naturwissenschaften*, 23:37–38.

Hahn, O. and Meinter, L. (1935b). Über die künstliche Umwandlung des Urans durch Neutronen (II mitteil.). *Naturwissenschafte*, 23:230–231.

Hahn, O. and Straßmann, F. (1939a). Nachweis der Entstehung aktiver Bariumisotope aus Uran und Thorium durch Neutronbestrahlung; Nachweis weiterer aktiver Bruchstücke bei der Uranspaltung. *Die Naturwissenschaften*, 27:89–95.

Hahn, O. and Straßmann, F. (1939b). Zur Frage nach der Existenz der 'trans-urane'. *Die Naturwissenschaften*, 27:451–453.

Herrmann, G. (1995). The discovery of nuclear fission—Good solid chemistry got things on the right track. *Radiochemistry Acta*, 70/71:51–67.

Krafft, F. (1981). *Im Schatten der Sensation: Leben und Wirken von Fritz Straßmann*. Verlag Chemie, Weinheim.

Krafft, F. (1983). Internal and external conditions for the discovery of nuclear fission by the berlin team. In Shea, 1983, pages 135–165.

Lakoff, G. (1987). *Women, Fire, and Dangerous Things. What Categories Reveal about the Mind*. University of Chicago Press, Chicago.

Meitner, L. (1934). Atomkern und periodisches System der Elemente. *Naturwissenschaften*, 22:733–739.

Meitner, L. (1936). Künstliche Umwandlungsprozesse beim Uran. In Bretscher, E., editor, *Kernphysik. Vorträge gehalten am Physikalischen Institut der Eidgenössischen Technischen Hochschule Zürich im Sommer 1936*, pages 24–42. Springer, Berlin.

Meitner, L. and Frisch, O. (1939). Disintegration of uranium by neutrons: a new type of nuclear reaction. *Nature*, 143:239–240.

Meitner, L. and Hahn, O. (1936). Neue Umwandlungsprozess bei Bestrahlung des Urans mit Neutronen. *Naturwissenschaften*, 24:158–159.

Meitner, L., Hahn, O., and Straßmann, F. (1937). Über die Umwandlungsreihen des Urans, die durch Neutronenbestrahlung erzeugt werden. *Z. Phys.*, 106:249–270.

Nersessian, N. and Andersen, H. (1997). Conceptual change and incommensurability: A cognitive-historical view. *Danish Yearbook of Philosophy*, 32:111–151.

Nickles, T. (1980). Can scientific constraints be violated rationally? In Nickles, T., editor, *Scientific Discovery, Logic, and Rationality*, pages 285–315. Reidel, Dordrecht.

Noddack, I. (1934a). Das Periodische System der Elemente und seine lücken. *Angew. Chem.*, 47:301–305.

Noddack, I. (1934b). Über das Element 93. *Angew. Chem.*, 47:653–655.

Seaborg, G. T. (1989). Nuclear fission and transuranic elements? 50 years ago. *J. Chem. Educ.*, 66:379–384.

Segré, E. (1970). *Enrico Fermi Physicist*. University of Chicago Press, Chicago.

Shea, W. R., editor (1983). *Otto Hahn and the Rise of Nuclear Physics*. Reidel, Dordrecht.

Stuewer, R. H. (1983). The nuclear electron hypothesis. In Shea, 1983, pages 19–68.

Stuewer, R. H. (1994). The origin of the liquid-drop model and the interpretation of nuclear fission. *Perspectives on Science*, 2:76–129.

Treumann, R. A. (1991). A post-fission perspective of the discovery of nuclear fission. *J. Gen. Phil. Sci.*, 22:143–153.

van Assche, P. (1988). The ignored discovery of the element z=43. *Nucl. Phys.*, A480:205–214.

von Grosse, A. (1934). The chemical properties of elements 93 and 94. *J. Am. Chem. Soc.*, 57:440–441.

von Grosse, A. and Agruss, M. (1934a). The chemistry of element 93 and Fermi's discovery. *Phys. Rev.*, 46:241.

von Grosse, A. and Agruss, M. (1934b). Fermi's element 93. *Nature*, 134:773.

von Weizsäcker, C. F. (1936). Metastabile Zustände der Atomkerne. *Die Naturwissenschaften*, 24:813–814.

von Weizsäcker, C. F. (1937). *Die Atomkerne*. Springer, Berlin.

Weart, S. (1983). The discovery of physics and a nuclear physics paradigm. In Shea, 1983, pages 91–133.

CONCEPTUAL COMPARISON AND CONCEPTUAL INNOVATION

Harold I. Brown
Department of Philosophy
Northern Illinois University
hibrown@niu.edu

The guiding idea of this paper is that discussions of comparisons of concepts across theories (individuals, historical periods, cultures) and the introduction of new concepts must be based on an account of how the content of concepts is determined. I will sketch a theory of concepts based on the work of Wilfrid Sellars, although with several modifications.[1] Then I will illustrate the application of this theory in two cases. First, I will compare the concepts of earth, water, and air as they appear in Aristotelian physics and in Galileo. Second, I will consider the concept of an isotope, an example of a new concept whose introduction is fairly well localized in the history of chemistry. There are two general conclusions that I want to draw from this discussion. First, a Sellarsian approach provides a specific set of tools for comparing concepts and introducing new concepts. Second, major conceptual change can take place while maintaining a great deal of continuity with existing conceptual resources.

One central theme of Sellars' theory of concepts is that concepts occur only as members of systems of interrelated concepts. At least part of every concept's content is determined by implications which hold between that concept and other concepts in the system. While holistic, this view should be read as a local holism. It does not require that all concepts link together into a single massive conceptual scheme. Rather, each of us deploys many different conceptual systems that have a variety of relations to each other. I have concepts that I use for thinking about poker, and some of these concepts have close ties to concepts that I use for thinking about other card games, and perhaps other games; but they have little connection with concepts that I use for thinking about

[1] See Brown, 1986 for a more detailed account.

J. Meheus and T. Nickles (eds.), *Models of Discovery and Creativity*,
DOI 10.1007/978-90-481-3421-2_2, © Springer Science + Business Media B.V. 2009

carpentry or plate tectonics. I also have two conceptual schemes for thinking about space and time—one from everyday experience and one from relativity theory. There are close and complex relations among the concepts in the two schemes, and there are good reasons for describing both as systems of space and time concepts.[2] Still, they can be treated as distinct conceptual schemes and I can shift from one to the other without confusing them.

Although implications are involved in determining the content of every concept, they are not always the complete story. Sellars distinguishes three types of conceptual systems. *First*, formal concepts—concepts of logic and pure mathematics—are wholly determined by the implications in which they play a substantive role.[3] There is no more or less involved in the concept of conjunction than is captured by the characteristic inferences that depend on this operator. Classical and intuitionist negation are similar, but not identical, concepts and their similarities and differences are completely determined by the implications in which they play an essential role.[4]

Second, there are descriptive concepts—these are among the most common concepts of everyday life and science. The content of these concepts is determined by a combination of implications and what Sellars calls "entry transitions". The idea is that there are paradigm instances of these concepts, and mastering a descriptive concept requires learning to recognize these cases. For example, to master the concept of a table I must learn conceptual relations such as that tables are furniture, manufactured objects, and commonly used for holding other objects at a convenient height; and I must also learn to spontaneously recognize typical tables as tables. If I have learned to parrot expressions such as "All tables are furniture" but cannot recognize a typical table, then I have not mastered this concept. Nor have I mastered the concept if I have learned to say "table" on encountering a typical table, but do not have the concepts of furniture or manufactured objects. We have here a key contrast with empiricist theories, which hold that mastery of a concept such as *red* requires only that I be able to correctly identify instances of red. On typical empiricist accounts, I might learn the concept red without acquiring any other color concepts, or the concept of a color, or indeed any other concepts at all. For Sellars, mastery of a concept always requires mastery of other concepts, but in the case of descriptive concepts more is required. Sellars describes the additional element as an

---

[2]See Sellars, 1973 and Sellars, 1974 for discussion of this and related examples.

[3]The point about a substantive role is required because an operator such as negation may be carried along through a series of implications that do not depend on this operator; these implications are not relevant to the content of this concept.

[4]These roles are not completely captured by the customary introduction and elimination rules. In accord with Sellars' overall approach, we should not think that the propositional operators can be specified one by one. For example, De Morgan's laws play an essential role in determining the classical concepts of negation, disjunction, and conjunction.

"entry *transition*" to underline a key point of his account: When we subsume an item under a concept we are making a transition from noticing that item into a conceptual system.

Third, there are normative concepts which have their content determined by a combination of implications and "departure transitions"—spontaneous moves from thinking about a concept to action in the world. To go from thinking of sitting on a chair to actually sitting would be one example of a departure transition, but the most important application of departure transitions occurs in Sellars' account of normative concepts. Sellars holds that normative concepts enjoin action, and that mastery of such concepts requires that I at least have a tendency to carry out this action.

It seems clear that we should add a *fourth* type of concept to this scheme since many concepts have both descriptive and normative aspects. Truth is a good example: saying that a proposition is true involves both a descriptive claim about that proposition and an injunction to believe and be prepared to act on that proposition. However, for the remainder of this paper I will be concerned only with descriptive concepts.

I now want to note two respects in which I am going to depart from Sellars' own practice. The first concerns a tendency, which he shares with many others, to use the terms "conceptual system" and "language" interchangeably, and to describe entry and departure transitions as moves between language and the world. One reason for avoiding this usage is that I do not want to make any assumptions about whether non-linguistic beings have concepts. In addition, this practice tends to limit our scope in discussing descriptive concepts. I want to pursue the thesis that whenever we have a body of beliefs about some subject matter, those beliefs are embodied in a system of descriptive concepts. One subject about which we have such beliefs is language. For example, a set of grammatical concepts is a conceptual system used to describe certain aspects of languages. When we recognize a word as a noun, we are making an entry transition from a bit of language into a system of grammatical concepts, and it is awkward, at best, to describe this as a transition from the world into language. To achieve the level of generality I wish to pursue, I will talk of "conceptual systems" and their "extra-systemic" subject matter—where that subject matter is external to the system of concepts that is used to describe it. I will take entry transitions to be transitions from a specific domain into the conceptual system we use to describe that domain.

My second departure from Sellars' practice is more significant, but the story is more complex. Sellars typically discusses concepts in psychological terms and the development of appropriate habits plays a central role in these discussions. Sellars' entry transitions are habits that take us unreflectively from noticing an item to thinking of a concept. In addition, where I have spoken of "implications" among the concepts in a system, Sellars talks of habitual "inferences". Further, when Sellars talks about mastering a concept, he is usually concerned with developing these habits. This psychological focus derives from Sellars' naturalistic view that concepts exist only in cognitive agents so that without such agents there would be no concepts. In addition, there are two main reasons for the emphasis on habits. First, Sellars is concerned with understanding concepts as tools by which we find our way around in the world, and this practical focus often requires that we respond to situations swiftly. We accomplish this by embodying our concepts in habits. Second, as Sellars points out (Sellars, 1963, p. 321), his holism generates a problem about how concept-learning ever gets started. Sellars holds that genuine mastery of a concept requires not only that we use the concept appropriately, but that in doing so we are obeying rules, formulated in a metalanguage, that govern the use of these concepts. Stated in his characteristic linguistic idiom, this suggests that in order to learn a language we must first learn the metalanguage, and this would seem to make it impossible for language learning to get started. Sellars' response is that—at least initially—we learn concepts in two stages. First, we develop habits that are reinforced in a social setting. Only later do we come to understand the rules that govern these habits and move to full competence in the use of these concepts.

This concern with habits is not relevant to the present paper. Sellars' second stage in concept learning involves a considerably more sophisticated mastery of concepts than is required when young children are first learning concepts, and we are working on this sophisticated level when we study concepts and propose new concepts. When we are engaged in these activities, we treat concepts as abstracted from whatever embodiments they have in habits. Indeed, when we contemplate conceptual change or examine the conceptual system of an abandoned scientific theory, we regularly master concepts without learning to apply them habitually. Sellars seem to recognize this point even while he is emphasizing the desirability of embodying concepts in habits. For example, he writes:

> suppose that 'ϕ' and 'ψ' are empirical constructs and that their conceptual meaning is constituted, as we have argued, by their role in a network of material (and formal) moves. Suppose that these moves do not include the move from 'x is ϕ' to 'x is ψ'. Now suppose that we begin to discover (using this frame) that many ϕ's are ψ and that we discover no exceptions. At this stage the sentence 'All ϕ's are ψ' looms as an 'hypothesis', *by which is meant that it has a problematical status with respect to the categories of explanation.* In terms of these categories we look to a resolution of this problematical situation along one of the following lines.
>
> (a) We discover that we can derive 'All ϕ's are ψ' from already accepted nomologicals. (Compare the development of early geometry.)
>
> (b) We discover that we can derive, 'If C, then all ϕ's are ψ' from already accepted nomologicals, where C is a circumstance we know to obtain.

> (c) We decide to adopt—*and teach ourselves* [italics mine]—the material move from 'x is ϕ' to 'x is ψ'. In other words, we accept 'All ϕ's are ψ' as an unconditionally assertable sentence of L, and reflect this decision by using the modal sentence 'ϕ's are *necessarily* 'ψ'. This constitutes, of course, an enrichment of the conceptual meanings of 'ϕ' and 'ψ'. (Sellars, 1963, p. 357)

Note the two main steps in Sellars' third case. First we conclude that all ϕ's are ψ and modify ϕ to include this condition; second, we undertake to adjust our habits so that we will spontaneously infer ψ from ϕ. But this second step requires that we already grasp the relevant concepts before we decide to teach ourselves the new inference.

In order to discuss concepts on this reflective level, we must make some changes at least in Sellars' terminology. One change is straightforward, and I have already made it: talking of implications rather than inferences. This change includes recognition that there may be more involved in a conceptual system than its users have so far recognized—a point that is implicit in Sellars' cases (a) and (b). Exploration of these implications is an important form of research and one possible source of reasons for engaging in conceptual change. Russell's discovery that classical set theory is inconsistent provides a striking illustration.

A more complex problem faces us in the case of entry transitions. The notion of an habitual move will be replaced by the requirement that the content of a descriptive concept include an account of the criteria for an item to be an instance of that concept. Such criteria will be internal to the conceptual system, rather than a direct link between that system and extra-systemic items, and will provide an account of how that link is to be established. For example, in discussing the concepts of Aristotelian physics we need an account of the cases that an Aristotelian would spontaneously identify as instances of, say, violent motion. The need for such criteria is also clear when we are introducing a new concept that may not have any instances. I will, however, continue to use Sellars' term "entry transition" for this aspect of descriptive concepts.[5]

This leaves Sellars' general thesis that concepts exist only as items in cognitive agents, with which I agree. An immediate consequence is that our psychol-

[5]This is only an introductory sketch; the full story is considerably more complex. For example, a charged particle may have a standard signature in a detector, but the reasons for believing that an uncharged particle passed through the detector may be just the absence of any charged particles in a particular context. Moreover, many cases require statistical analysis to determine if a particular particle passed through the detector. Consider the concept of the top quark. This concept is well understood by physicists, and they have evidence that the concept is instantiated. But this evidence does not include any single detector output indicating that the particle occurred. Rather, it consists of a body of data for which the occurrence of a top quark is one of a set of possible explanations, plus an argument to show that the probability that none of these cases involved a top quark is incredibly low. Most of Sellars' examples concern simple observables and the extra-systemic side of an entry transition is usually an instance of the concept. But Sellars is aware that these more complex cases exist (e.g., Sellars, 1963, p. 316) even though he does not consider any in detail.

ogy and biology provide a major constraint on the acceptability of a theory of concepts. A theory that attributes to concepts properties that cannot be embodied in human biology and psychology cannot provide either a correct account of human conceptual development or a set of recommendations for how we should endeavor to introduce new concepts. Nevertheless, this issue can be left aside in the present context. For purposes of analysis, concepts can be treated as abstract structures apart from their actual embodiments—although this approach leaves open the possibility that our results may be undermined by evidence from psychology or biology.[6]

I turn next to a central and controversial feature of Sellars' account of descriptive concepts. As was indicated in the earlier quote, Sellars holds that when we firmly accept the empirical generalization "All A are B" we build the implication from A to B into our concept of A. The generalization, which is in the metalanguage governing this system of concepts, now functions as a material rule which, along with formal rules, licenses the implications associated with A. Sellars says surprisingly little about the analytic/synthetic distinction, but he does deny that this is a distinction between propositions that are determinative of the content of concepts and those that are not.[7] In effect, Sellars holds that the concept of an A embodies all of our firm beliefs about A's and that we change our concepts when these beliefs change.[8] The view that material rules enter into the content of concepts has several immediate consequences: conceptual change is more common than many philosophers take it to be, and the dividing

[6]I have written this paper so that it is neutral between a necessary-and-sufficient-conditions view of concepts and a view of concepts as having open texture. The former thesis requires that examples of conceptual change be viewed as cases in which one concept is replaced by a different concept, but this does not alter the point that we can still explore similarities and differences between a concept and its replacement. Still, an open-textured view seems more appropriate for a naturalistic approach since it allows for the idea that we often introduce concepts in response to current needs without thinking through many of their ramifications until a need to do so arises.

[7]This amounts to the proposal that we make certain changes in the traditional system of epistemic concepts. Quine's rejection of the distinction is another proposal of this kind. See Brown, 1991 for further discussion.

[8]Sellars has many reasons for adopting this approach. 1. One reason derives from his view that concepts guide our action in the world. We build our firm beliefs about items into the associated concepts in order to assure that we respond appropriately to items when we encounter them. 2. The approach provides a way of absorbing a point that has been argued, in different ways, from Kant to Kuhn and beyond: Fruitful scientific research requires the acceptance of propositions that are not analytic but that are protected from empirical refutation, at least for a time. These propositions play a central role in providing the conceptual framework within which research takes place. In Sellars' version, this results in propositions that are non-analytic, but true *ex vi terminorum* (cf. Sellars, 1963, ch. 10). 3. The view is a central part of Sellars' project of analyzing causal claims as metalinguistic claims about our descriptive concepts. He proposes that we reformulate the problem of induction as a concern with our decisions to accept *specific* empirical correlations and then build them into our concepts. As a result, accepting a material rule is equivalent to believing a causal necessity (cf. Sellars, 1958). 4. The approach also provides the basis for an account of how we can learn concepts in a piecemeal fashion, elaborating a concept as we learn more about what features are included in it in our society. 5. Most important, for present purposes, we will see that the view is an integral part of an account of how we can introduce new concepts by building on available concepts, and learn older concepts by backtracking from current concepts to those from which they were historically derived.

line between change of concept and change of belief is extremely vague. It also follows that conceptual diversity is more common in a community than it is often taken to be by philosophers. But this diversity does not generate massive failures of communication because the differences between concepts in a population may be small. Differences between two descriptive concepts can consist of differences in accepted implications, entry transitions, or both. At the same time, we have here a basis for introducing new concepts by making systematic changes in entry transitions and implications of existing concepts.

I now want to enrich the Sellarsian theory of concepts by extending the scope of an idea that Sellars deploys only in a specific case. Sellars argues that new entities are introduced by analogy with familiar entities: the new entity is conceived of as identical with familiar entities in some respects, but having additional properties, or lacking properties of those familiar entities. For example, molecules can be introduced as very small spheres, like billiard balls, but lacking color and temperature while being capable of completely elastic collisions. Moreover, such analogies are not limited to first-order properties. The common analogy between successive moments of time and points on a directed line is based on properties of the ordering relation. The introduction of a new entity is always accompanied by a "metalinguistic commentary" in which we explain the identities and differences between the new entity and whatever provides the basis of the analogy (Sellars, 1963, ch. 5; Sellars, 1965).

Now the introduction of a new entity amounts to the introduction of a new concept, and Sellars is here describing a process by which we introduce new concepts by analogy with available concepts. But analogous concepts are just concepts that are the same in some respects and different in others, and there is no reason why the process need be restricted to cases concerning entities. New concepts of any kind can be introduced by such analogies with existing concepts. In the case of descriptive concepts these analogies can involve identities and differences in implications and in entry transitions. We may also compare concepts from competing or successive scientific theories by exploring such analogies.[9] Such discussions are always metalinguistic, and I will take Sellars' notion of a metalinguistic commentary as a prototype for all discussions of concepts. When we are carrying out such discussions we have available all of the language and concepts that are required to achieve this level of cognitive sophistication.

Once we look at concepts from this metalinguistic perspective, another point that Sellars alludes to from time to time comes into focus. Each of our scientific concepts has been introduced to do a specific cognitive job. Indeed, one reason for introducing a new concept is that we come to recognize the need for a

[9] We may even be able to approach concepts from another historical or contemporary human society by mapping out analogies with our own concepts.

cognitive job that was previously not recognized; Newton's distinction between weight and mass provides one example. At the same time, we drop concepts when we reject the cognitive jobs that they had been introduced to carry out. Rejection of the traditional distinction between a terrestrial and a celestial realm is an example. Now Sellars never develops this idea or integrates it into his overall theory of concepts. Strictly speaking, when Sellars writes of a "conceptual role" he means just the appropriate combination of implications and entry (or departure) transitions.[10] I suspect that this is a direct result of his focus on the active use of concepts and their embodiment in habits, since to master a concept in use we need learn only its implications and transitions. But once we move to reflective discussions of our concepts, consideration of the role a concept plays in our cognitive economy provides a key part of an account of that concept, and comparisons of conceptual roles provide an additional dimension for comparing conceptual systems.

The upshot of this sketch is that a Sellarsian approach provides us with three specific dimensions to work along when we are analyzing a descriptive concept, proposing new concepts, and comparing concepts from different scientific theories: implications, entry transitions, and what I will call "conceptual roles". I want to illustrate the power of this approach by applying it, first, to the comparison of a set of concepts from Galileo's physics with a related set from Aristotle, and then by examining the introduction of the concept of an isotope.

An essential part of Galileo's dynamical theory, as developed in his *Dialogue on the Two Chief World Systems*, is a distinction between the elements of earth, water, and air, where these are characterized by their dynamical properties.[11] I will sketch Galileo's account of these concepts and then use the Sellarsian approach to compare them with the versions that occur in Aristotle's physics. Earth, according to Galileo, is characterized by three kinds of natural motion plus the ability to sustain an impressed motion. A *natural motion* is a motion that an object pursues when not constrained or acted on by an external force. One of these is the motion of an object to its natural place. This accounts for the fall of unsupported objects and is worthy of further exploration, but I will focus here on the two additional natural motions that Galileo introduces.[12] There are two such motions: a daily motion around the center of the planet and an annual motion around the sun. Thus Galileo requires no special explanation for the daily rotation and annual revolution of the planet: these are simply the

[10]In one place Sellars seems to explicitly reject the notion of a conceptual role that I am introducing. Discussing the German name *Sokrates*, Sellars writes: "One is tempted to say that the function in question is that of being used to refer to a certain Greek philosopher. But it is a mistake to tie the semantical concept of a reference too closely to referring as an illocutionary act" (Sellars, 1974, p. 428).

[11]Galileo expresses doubt that fire is an element. See Brown, 1976 for further details and references.

[12]While Galileo holds that fall occurs because an object is moving to its natural place, his account of natural place is significantly different from Aristotle's.

natural motions of the predominant element in its composition. This doctrine of natural motion provides the basis for Galileo's response to several standard anti-Copernican arguments. For example, in the case of the tower argument Galileo argues that the fact that a rock dropped from the top of a tower lands at the base is compatible with a rotating earth because the rock, an earthy object, engages in the natural daily rotation of the earth, and thus maintains its relative location with respect to the tower as it falls. Nor does it follow from Copernicanism that an arrow shot vertically would land far to the west of the archer because of the earth's annual motion, since the arrow is another earthy object which shares that natural motion.

Impressed motion occurs when an object is pushed into some non-natural motion by an external force; projectile motion is the key case. Earthy objects sustain an impressed motion, and this is the basis for Galileo's prediction that a rock dropped from top of the mast of a moving ship would land at the foot of the mast, not at the rear of the ship, as Aristotelians predicted.[13]

Now consider water. According to Galileo, water does not share the natural motion of the earth, but does sustain an impressed motion. This is the dynamical basis for Galileo's theory of the tides, which he considered a particularly powerful argument for the motion of the earth. Because of the double motion of the earth, water confined in an ocean basin is subject to a pair of continually changing impressed motions. The water tends to sustain these impressed motions, and tides result from the sloshing of the water in its basin.

Air does not share the natural motion of the earth and does not sustain an impressed motion. There are two anti-Copernican arguments that concern the air: if the earth rotates from west to east, we should experience a continual wind blowing from east to west; and, as a result of the earth's annual motion, the earth should leave the air behind. Galileo replies that the air is drawn along with the earth because it is trapped by the roughness of the earth, and also because it is mixed with "earthy vapors". But while this will explain why we do not lose our atmosphere or experience a constant wind over land, Galileo contends that we do find exactly the predicted wind over the oceans. An apparent counter-instance to Copernican astronomy is thus turned into a confirmation.

Let us compare Galileo's concepts of earth, water, and air with their Aristotelian counterparts. First, these concepts play similar roles in the two frame-

[13] Galileo replies to the major physical arguments against a moving earth by appeal to natural motion, not impressed motion. This has an intriguing consequence for the ship experiment—which Galileo did not do and apparently had no interest in actually doing. If Galileo is correct and the object falls at the foot of the mast, the experiment supports the Copernican view since it would show that an object is capable of engaging in multiple motions simultaneously. If this is the case for an impressed motion, the point holds *a fortiori* for natural motion. If the rock were to fall at the rear of the ship we would have an empirical challenge to Galileo's account of impressed motion, but no significant argument against the Copernican view since Galileo's defense of that position depends only on natural motion.

works. In both cases, the concepts pick out basic kinds of entities in the terrestrial world, and these entities are characterized in terms of their dynamical properties. By the same token, the concept of natural motion plays similar roles in the two frameworks—it captures how a type of object moves when it is not being forced or restrained. Second, with regard to entry transitions consider, first, identifications of a sample as earth, water, or air. Here we find complete agreement between the two frameworks. However, the story becomes a bit more complex when we turn to the concept of natural motion. In the case of earth, Galileo includes two kinds of natural motion that have no place in an Aristotelian framework. But these new natural notions pertain only to earth, so Galileo's account of natural motions for air and water would presumably agree with Aristotle's. On the other hand, there would be no Galilean entry transitions to the concept of violent motion since this concept, which Aristotle defines as a contrary to natural motion, does not exist in Galileo's conceptual scheme.

The deepest differences between the two frameworks come out when we look at implications. For example, since Galileo drops the concept of violent motion, the implications associated with this concept in the Aristotelian framework vanish. This is a particularly important change because it is Aristotle's definition of natural and violent motion as contraries that yields the supposedly a priori truth that no object can be moving simultaneously in a horizontal and a vertical direction. Dropping this fundamental contrast opens up logical space for multiple kinds of natural motions, for the simultaneous occurrence of natural and impressed motion, and for multiple impressed motions in a single object at the same time. This is only a cursory discussion, but it is sufficient to support the thesis that when we examine Galileo's system of dynamical concepts using the tools of our Sellarsian theory of concepts, we can see that at least parts of this system constitute a systematic alteration of Aristotle's dynamical concepts. I urge that this kind of detailed comparison is considerably more informative than attempts to give simple "Yes" or "No" answers to such questions as whether the two systems are commensurable.

Now consider the concept of an isotope, a new concept that was introduced into chemistry when it became clear that a new conceptual role was required. Here is the relevant background. The thesis that a characteristic weight is the defining feature of each chemical element was central to nineteenth-century chemistry. This view had been introduced by Dalton, was embodied in Prout's thesis that each element is compounded out of hydrogen atoms, and provided a major part of the conceptual basis for locating elements on the periodic table. Anomalies appeared throughout the century, so that by 1886 Crookes put forward the "audacious" but testable speculation that the weight standardly associated with an element was that of the majority of its atoms, and that some might have slightly different weights (Bruzzaniti and Robotti, 1989, p. 309). Still, the prevailing view was that variant atomic weights associated with a specific element indicated failures of chemical analysis.[14] The proposal that chemically identical pure samples could differ in weight involved a deep change in chemical thought. As Soddy noted, it undercut a central research project of nineteenth-century chemistry:

> There is something, surely, akin to if not transcending tragedy in the fate that has overtaken the life work of that distinguished galaxy of nineteenth-century chemists, rightly revered by their contemporaries as representing the crown and perfection of accurate scientific measurement. Their hard-won results, for the moment at least, appears as of as little interest and significance as the determination of the average weight of a collection of bottles, some of them full and some of them more or less empty. (Soddy, 1932, p. 50)

The main impetus for introducing this change came from the recently discovered phenomenon of radioactivity. By 1913, through the work of several researchers, it had become clear that transformations occurred in which an element emitted an alpha particle and two beta particles (in any order). This left its slot in the periodic table unchanged while its weight dropped by four units (Fajans, 1913; Soddy, 1913b).[15] This led Soddy to propose a new basis for locating elements on the periodic table.[16] He believed that the nucleus contained both electrons and protons and that the difference between these—the "intra-atomic charge"—provided the proper criterion.[17] *Isotope* is a new concept; let us consider its introduction from our Sellarsian perspective.

The concept of an isotope marks a new conceptual role, one which was not only unnecessary in the pre-existing system of chemical concepts, but actually precluded. Previously, the concept used to describe each element implied a characteristic weight, and while not every weight is an atomic weight, every *atomic* weight implied a specific element. Both of these implications were dropped when the new role was introduced: now an element could have different weights, and different elements could have the same weight. As a result of these changes, implications between a weight and location on the periodic table were dropped and replaced by a new mutual implication between location on the

[14]In other words, the thesis that elements are characterized by their weight played the role of a guiding assumption (cf. Laudan et al., 1986): A variety of chemical tests were used to identify elements, and samples initially identified as the same element could exhibit different atomic weights, but this was not considered evidence against the thesis that elements are characterized by weight. Instead, such cases were interpreted as evidence that impurities were still present.

[15]Throughout this period it was assumed that electrons make no significant contribution to an element's weight, although it was recognized that electrons do have mass. Thus beta decay was treated as involving no change of weight.

[16]As Soddy noted, the same proposal was made slightly earlier by van den Broek (1913), although his concerns were different: he was attempting to bring the periodic table into accord with the thesis that all elements were built up out of halves of alpha particles.

[17]Clearly, the concept of an intra-atomic charge is not the same as the modern concept of atomic number because it assumes a view of the nucleus that is now rejected and, as a result, it is calculated in a way that makes no contemporary sense.

periodic table and net nuclear charge. Yet the set of implications that constituted most of the existing body of chemical knowledge stood unaltered. Indeed, the arrangement of elements in the periodic table was not affected, even while the conceptual basis of this ordering was undercut. In addition, all results of standard chemical and spectroscopic analyses remained unchanged. Even the vast majority of implications among elements and physical properties endured, although those explicitly involving considerations of atomic weight, such as density and diffusion rates, had to be reconsidered (cf. Soddy, 1932, p. 44).

These changes in implications are directly reflected in changes in entry transitions, i.e., in the tests for specific elements and isotopes. Measurements of weight were greatly reduced in significance, and new tests were needed to distinguish isotopes of an element. In effect, these required the ability to detect small weight differences in chemically indistinguishable samples; the most important technique was soon embodied in Aston's mass spectrograph. In addition, the newly discovered property of the half-life provided a means of recognizing different isotopes of an element—as well as a new means of distinguishing among radioactive elements. These were radical changes, yet it is striking how much accepted chemical practice and knowledge remained unchanged even while their foundation was being radically restructured.[18]

These examples bring us to the two conclusions announced at the beginning of this paper. First, the Sellarsian theory of concepts provides a systematic approach to the analysis of conceptual innovation and conceptual change, in particular, to sorting out what changed and what remained essentially the same in specific cases. It also provides a guide to the process of introducing new concepts. Second, and more generally, when we approach specific changes from this perspective, we see clearly that change is not an all-or-nothing phenomenon, and that radical conceptual change in a field is quite compatible with a great deal of stability. These stable elements provide the basis for carrying out conceptual innovation in a coherent manner and for debating the merits of a new framework.

References

Brown, H. (1976). Galileo, the elements, and the tides. *Studies in History and Philosophy of Science*, 7:337–351.

Brown, H. (1986). Sellars, concepts and conceptual change. *Synthese*, 68:275–307.

Brown, H. (1991). Epistemic concepts. *Inquiry*, 34:323–351.

Bruzzaniti, G. and Robotti, N. (1989). The affirmation of the concept of isotopy and the birth of mass spectrography. *Archives Internationales D'Histoire des Sciences*, 39:309–334.

Fajans, K. (1913). The placing of the radioelements in the periodic system. In Romer, 1970, pages 205–219.

[18]The concept of an isotope is just one locus of this rethinking. The research leading to this new concept came to a head in 1913, just a few months before Bohr's new theory of the atom was published.

Laudan, L., Donovan, A., Laudan, R., Barker, P., Brown, H., Leplin, J., Thagard, P., and Wykstra, S. (1986). Scientific change: Philosophical models and historical research. *Synthese*, 69:141–223.

Romer, A., editor (1970). *Radiochemistry and the Discovery of Isotopes.* Dover Publication, New York.

Sellars, W. (1958). Counterfactuals, dispositions, and the causal modalities. In Feigl, H., Scriven, M., and Maxwell, G., editors, *Concepts, Theories, and the Mind-Body Problem*, pages 225–308. University of Minnesota Press, Minneapolis.

Sellars, W. (1963). *Science, Perception and Reality.* Humanities Press, New York.

Sellars, W. (1965). Scientific realism or irenic instrumentalism: A critique of Nagel and Feyerabend on theoretical explanation. In Cohen, R. and Wartofsky, M., editors, *Boston Studies in the Philosophy of Science,* Volume II, pages 171–204. Reidel, Dordrecht.

Sellars, W. (1973). Conceptual change. In Pearce, G. and Maynard, P., editors, *Conceptual Change*, pages 77–93. Reidel, Dordrecht.

Sellars, W. (1974). Meaning as functional classification. *Synthese*, 27:417–437.

Soddy, F. (1913a). Intra-atomic charge. *Nature*, 92:399–400.

Soddy, F. (1913b). The radio-elements and the periodic law. In Romer, 1970, pages 219–228.

Soddy, F. (1932). *The Interpretation of the Atom.* John Murray, London.

van den Broek, A. (1913). Intra-atomic charge. *Nature*, 92:372–373.

DISCOVERING MECHANISMS IN MOLECULAR BIOLOGY

Finding and Fixing Incompleteness and Incorrectness

Lindley Darden
Department of Philosophy
University of Maryland, College Park
darden@umd.edu

Abstract The discovery of mechanisms occurs in cycles of generation, evaluation and revision of hypothesized mechanism sketches and schemata. A new analysis of the concept of a mechanism by Machamer, Darden and Craver (2000) points to hitherto unexplored aspects of mechanism discovery. Incomplete sketches have their black boxes filled as research proceeds. When an anomaly requires revision of a hypothesized mechanism schema, different entities and/or activities may be proposed to fulfill a functional role in the mechanism. Temporal and compositional anomalies point to the types of revisions needed. Examples of hereditary mechanisms from Mendelian genetics and molecular biology illustrate reasoning in such on-going discovery processes. We need no longer debate whether there is a logic of discovery of theories; instead we can investigate reasoning strategies for the generation, evaluation and revision of mechanism schemata.

1. Introduction

This paper brings together two lines of research in history and philosophy of science: research on discovery and research on mechanisms. Thinking about mechanisms provides insights into the discovery process. This paper will focus on the way thinking about mechanisms aids in identifying and removing two failings: incompleteness and incorrectness.

Discovery is to be viewed as an extended process. In *Theory Change in Science*: *Strategies from Mendelian Genetics* (Darden, 1991), I discussed an extended discovery episode, the discovery of the theory of the gene. Scientific reasoning, I argued, should be viewed as problem solving, and, further, an

J. Meheus and T. Nickles (eds.), *Models of Discovery and Creativity*,
DOI 10.1007/978-90-481-3421-2_3,

important task is to find problem-solving strategies. Also, scientific discovery should be viewed as an extended process that occurs in cycles of generation, evaluation, and revision. Close relations among generation, evaluation and revision show that the old philosophical distinction between discovery and justification is not a useful one (Darden, 1991, Chs. 2, 15).

During hypothesis generation, the nature of the product guides the process of discovery. Knowledge that what is to be constructed is a representation of a mechanism provides constraints and guidance during generation that are not available if one merely says that one wishes to discover a theory. More or fewer constraints can be put into the generation process which will then affect the evaluation process. One would, ideally, like to have a sufficiently constrained problem so that the entire hypothesis space could be constructed, or, failing that ideal, at least have reasoning methods so that the most plausible hypotheses are generated. When there are fewer constraints on generation, then more subsequent evaluation is required. If we know nothing about the generation process and how thorough the search for plausible alternatives has been, then evaluation of any one candidate hypothesis may not be a reliable way of finding the best one. Thus, generation and evaluation are tightly linked processes.

Rarely is generation sufficiently constrained that revision is not needed. Modular subcomponents of a complex hypothesis may be individuated and separately revised when evaluation strategies indicate a type of failure has occurred (Darden, 1991).

During evaluation of a hypothesized mechanism, two possible kinds of failure, among many, are judgments of incompleteness and incorrectness. Incompleteness is a black box, a stage of the mechanism that has yet to be illuminated. Incorrectness, on the other hand, shows that a hypothesized mechanism component does not properly fit. Incorrectness is often detected as a result of finding an anomaly. Diagrams of hypothesized mechanisms can aid revision, both to remove black box incompleteness and to resolve anomalies. Diagrams aid localization of the failure, and they indicate the organizational context within which a solution must fit.

The following sections are organized as follows. First will be a discussion of previous work characterizing mechanisms. Then examples from Mendelian genetics will illustrate black box incompleteness, which was removed by discoveries in molecular biology. Finally, examples from the discovery of components of the mechanism of protein synthesis will illustrate how anomaly-driven revision occurs and how thinking about mechanisms guides those reasoning processes. Thus, the discovery of hereditary mechanisms illustrates the way thinking about mechanisms aids in understanding reasoning in discovery in an extended period of mechanism discovery, spanning Mendelian genetics and molecular biology, from the beginning to the last half of the twentieth century.

2. Characterization of Mechanisms

A number of philosophers of science have noted the importance of mechanisms, including Wimsatt (1972), Brandon (1985), Bechtel and Richardson (1993), Burian (1996), Glennan (1996), and Thagard (1999). In recent work Peter Machamer, Carl Craver, and I characterize mechanisms in the following way:

> Mechanisms are entities and activities organized such that they are productive of regular changes from start or set-up to finish or termination conditions. (Machamer et al., 2000, p. 3)

Types of entities include ions, macromolecules (such as proteins and the nucleic acids, DNA and RNA), cellular structures (such as chromosomes), and cells (such as gametes). Types of activities include geometrico-mechanical activities, such as lock and key docking of an enzyme and its substrate, electrochemical activities, such as strong covalent bonding and weak hydrogen bonding, and cell fusion, such as the joining of egg and sperm during fertilization.

Mechanisms are made of components that work together to do something. The entities and activities are organized in stages with *productive continuity* from beginning to end. That is, each stage must give rise to, allow, drive, or make the next stage (Darden and Craver, 2002). One goal in discovering a mechanism is to reveal the mechanism's productive continuity. Any breaks in our understanding of that productive continuity constitute incompleteness, a black box yet to be illuminated. Each stage must be appropriately related to the next. Such relations can take many forms. Two examples are compositional relations between entities and temporal relations between successive stages in the mechanism. Compositional and temporal constraints are two among many constraints guiding mechanism discovery. (See the list in Table 1; only two will be discussed here, componency and temporal constraints.) Knowing what constraints an adequate mechanism description must satisfy guides generation of alternative hypotheses. If constraints were not used in the generation process, then they must be imposed in the evaluation process and they allow the detection of types of failures. For example, when a constraint is violated, then an anomaly results and revision is required. Once again we see the tight connections between generation, evaluation, and revision in the discovery of a mechanism.

Mechanism schemata (Skipper, 1999; Machamer et al., 2000) are abstract frameworks for mechanisms. They contain place holders for the components of the mechanism and they indicate, with variable degrees of abstraction, how the components are organized. These place holders may characterize a component's role (Craver, 2001) in the mechanism and show the context into which it must fit.

Many discoveries in biology involve discovering a mechanism schema. The general knowledge in molecular biology, for example, can be viewed as con-

Table 1. Constraints on the Organization of Mechanisms (from Craver and Darden, 2001, p. 134). This paper discusses the underlined componency and temporal constraints.

Character of phenomenon
Componency Constraints
Entities and activities
Modules
Spatial Constraints
Compartmentalization
Localization
Connectivity
Structural
Orientation
Temporal Constraints
Order
Rate
Duration
Frequency
Hierarchical Constraints
Integration of levels

sisting of knowledge of mechanism schemata. So far as I know, the phrase "theory of molecular biology" is not used. Instead, the general knowledge in the field is knowledge of a set of related mechanisms. For example,

$$\text{DNA} \rightarrow \text{RNA} \rightarrow \text{protein}$$

is a mechanism schema for the mechanism of protein synthesis. Other schemas are found in molecular biology for DNA replication and regulation of gene expression. (A similar point about the importance of mechanisms in molecular biology was made by Burian (1996).)

Mechanism schemata play the roles usually attributed to theories: they abstractly encapsulate general knowledge; they have varying scopes of applicability; they may be instantiated to provide explanations or predictions of particular phenomena. We need no longer debate whether there is a logic of discovery of theories; instead we can investigate reasoning strategies for the generation, evaluation and revision of mechanism schemata.

Discovering a mechanism involves constructing a schema. As we will see, diagrams are often employed to depict graphically the schematic organization of mechanisms. A mechanism sketch is an incomplete schema, with black boxes that cannot yet be properly filled (Machamer et al., 2000). In contrast to a sketch, a complete schema has place holders for all the working entities and their activities and filling instructions for how to instantiate the schema with particular entities and activities. When the schema is instantiated, it depicts a productively continuous mechanism from beginning to end, with no unfilled gaps.

3. Revision of Incomplete Schemata

There are numerous strategies for generation, evaluation, and revision of mechanism schemata. The generation strategies of schema instantiation, modular subassembly, and forward/backward chaining are discussed elsewhere (Darden, 2002). Craver laid out experimental strategies for testing hypothesized mechanisms with top down and bottom up experiments; he also showed how proposed mechanisms are evaluated by how well they fit into a hierarchy of nested mechanisms (Craver, 2002).

The focus in this paper is not on generation or evaluation strategies, but on strategies for revision. Given a puzzling phenomenon, one may be able to draw a rough sketch of a hypothesized mechanism that produces it. Sketches have black boxes indicating incompleteness. That incompleteness can be of two kinds. First, a completely unilluminated black box does not yet have a role specified for what is to fill it. For example, given the phenomenon of the partial resemblance of an offspring to its parent, a sketch for a hereditary mechanism puts the parent at the beginning and the child at the end. The middle is a completely unilluminated black box prior to the nineteenth century. Second, at a later stage, a more illuminated black box has a role specified, but its filler has not yet been found. (These two kinds of black boxes are discussed in Craver, 2001; Darden and Craver, 2002.) Work in Mendelian genetics produced the inference that genes segregated and independently assorted during the formation of parental gametes. Those black boxes in the hereditary mechanisms were filled with the discovery that, as T. H. Morgan put it, "the chromosomes furnish exactly the kind of mechanism" that Mendelism calls for (Morgan et al., 1915, p. viii).

Filling black boxes demands finding the working entities that act in the mechanism in such a way as to fulfill the hypothesized role. (The concept of a working entity is discussed in Darden, 2005.) Interestingly, the genes themselves are not the working entities in segregation and independent assortment. It is the chromosomes, the wholes of which the genes are parts, that do the work. Hence, one does not always decompose an entity to find the mechanism by which it operates (cf. characterizations of mechanisms by Glennan (1996) and Thagard (1999)). Sometimes one needs to find a larger working entity to find what its parts are doing. Some of the parts are just along for the ride.

Figure 1 illustrates the mechanisms of segregation and independent assortment. That figure shows that the chromosomes are the working entities in the mechanisms of both independent assortment, as chromosomes line up randomly

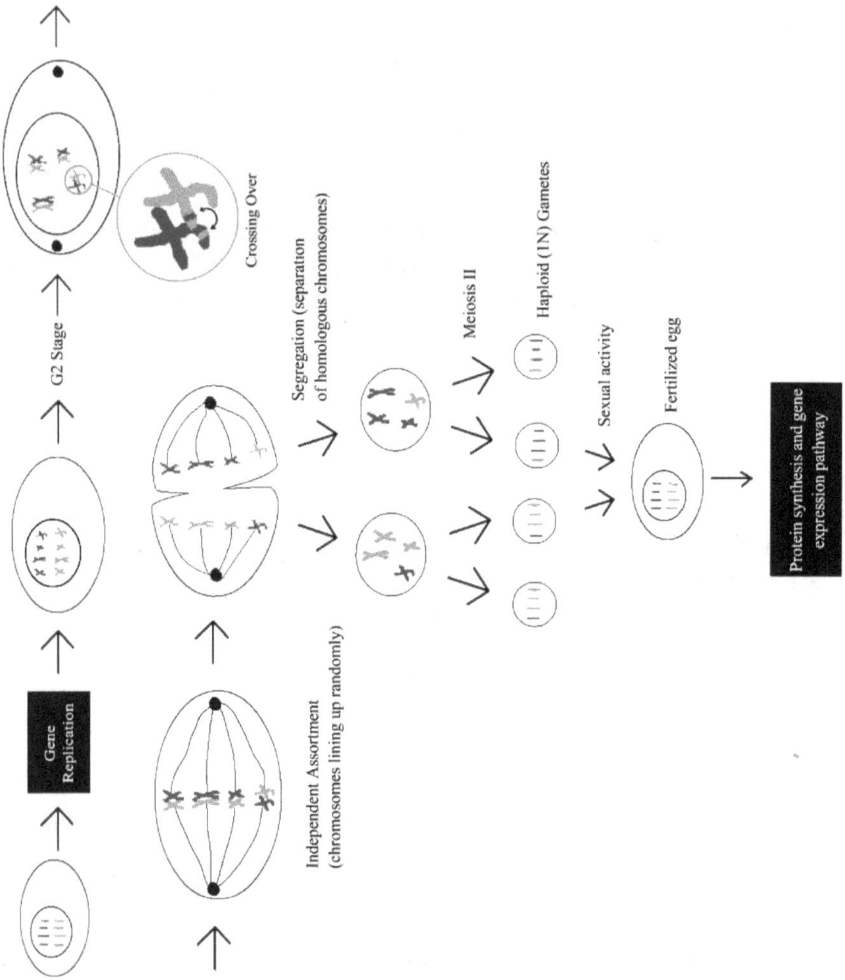

Figure 1. Mechanisms of Heredity

along the equatorial plate, and, subsequently, segregation, as chromosomes are pulled apart into separate cells. But further black boxes were left unilluminated by Mendelian and cytological techniques. How genes replicate was unfilled (see Figure 1, the black box labeled "gene replication"). Gene replication occurs before segregation and independent assortment. It was left unilluminated by Mendelian genetics and was only filled after the discovery of the double helix structure of DNA. As Watson and Crick (1953) noted, the structure immediately suggested a copying mechanism, with the unwinding of the two strands of the helix and the hydrogen bonding to new bases to produce two new helices. The entire DNA double helix is the working entity in gene replication. So, again in the DNA synthesis mechanism, we see it is not the individual genes themselves that are working entities in the mechanism; they are just along for the ride. Finding the appropriate working entities is an important step in discovering mechanisms.

Only in illuminating the black box of gene expression do the genes become working entities (see Figure 1, the black box labeled "Protein synthesis and gene expression pathway"). A first step in illuminating that black box was finding the role of genes in protein synthesis. The genes are segments of DNA that are transcribed into a template RNA, which then provides the order of the amino acids in a protein. (For discussion of that extended discovery episode see Judson, 1996; Rheinberger, 1997; Darden and Craver, 2002.)

Thus, the relations between Mendelian genetics and molecular biology, which have long puzzled philosophers, are to be understood in terms of temporally related mechanisms (Darden, 2005). Molecular biology illuminated the black boxes at earlier and later stages in the hereditary mechanisms that occur before and after the chromosomal mechanisms, as illustrated in Figure 1. When the chromosomes are unpacked for replication then the working entities are the entire DNA molecules in each chromosome. Later, during gene expression, the working entities are segments of DNA molecules, plus the other entities in the mechanism of protein synthesis and gene regulation.

The search to fill black boxes should not always be carried to the lowest size level. Instead, the search should be for the working entities of whatever size to appropriately fill the black box, that is, entities with activities that fill the roles required for that module. In hereditary mechanisms, the working entities differ in size from whole organisms that mate, to cells that combine at fertilization, to chromosomes that independently assort and segregate, to entire DNA molecules that replicate, to sections of DNA that are transcribed during protein synthesis. Entities often have to be of a certain size, having a certain structure, in order to carry out their roles in mechanisms.

Filling black boxes involves, first, specifying the roles to be filled and then finding the appropriate working entities and their activities to fill those roles.

Yet, even after hypothesized role fillers have been suggested, revision may be needed. Empirical tests often reveal anomalies.

4. Revision of Incorrect Schemata

In contrast to the black box incompleteness of mechanism sketches, incorrectness indicates that a hypothesized portion of a schema needs revision. Resolution of anomalies often drives scientific change (Darden, 1991, 1992). Anomaly resolution can be viewed as, first, a diagnostic reasoning process to localize where a failure has occurred, and, then, second, a redesign process, to propose an improved mechanism schema.

A diagram of a hypothesized mechanism schema identifies the stages of the mechanism and the place holders at each stage. These stages then become candidate localizations during anomaly resolution. Philosophers of science have been much too pessimistic about the ability to localize failures (e.g., Laudan, 1977). At least sometimes, anomalies can be localized quite specifically. Once an anomaly is localized in a stage, then the resolution may require a change before the stage, in the component entities and activities of the stage itself, or after the stage.

A particular type of anomaly will indicate that a particular type of change is needed. In this paper, only two types of anomalies will occupy us: a compositional anomaly and a temporal anomaly. A *compositional anomaly* indicates that a proposed entity does not have the required composition to play its assumed role in the mechanism. Alternatively, one kind of *temporal anomaly* results when something occurs more quickly than expected. Such temporal anomalies require activities that can occur more rapidly. (Again, see Table 1 for other constraints that could produce other kinds of anomalies when violated.) Both compositional and temporal anomalies indicated the need for a change during the historical development of our understanding of the mechanism of protein synthesis.

The black box of gene expression was opened by molecular biologists after the 1953 discovery of the double helix of DNA. When Watson (1968) put the sketch DNA → RNA → protein above his desk in 1952, the role of the RNA was a black box. This mechanism sketch of protein synthesis underwent numerous changes (Darden and Craver, 2002). Only one revision will occupy us here: the change in the view of the nature of the RNA template that carries genetic information from the DNA in the nucleus to the cytoplasm, where proteins are synthesized. By about 1955 it was thought that the ribosome filled the role of the template for carrying information from the DNA for the formation of proteins. (The ribosomes are particles in the cytoplasm of cells composed of RNA and proteins.) By 1961 that role was played instead by messenger RNA.

The resolution of two anomalies produced that change in the hypothesized mechanism.

These changes may be represented in these diagrams:

DNA → template → protein

DNA → ribosome → protein

DNA → messenger RNA → protein

By about 1955, the role of the template in protein synthesis was thought to be played by the ribosome. However, in the late 1950s, anomalies began to emerge for the view of the ribosome as carrying the coded sequence for ordering the amino acids. In 1958, two Russians, Belozerskii and Spirin showed: "the DNA of different microorganisms had greatly different base ratios... The base composition of the total RNA of these same organisms hardly varied at all..." Similar results had been reported by another group (Crick, 1959, pp. 35-36). Thus, replication of the result was important in taking it seriously as an anomaly for the prevailing view at the time.

Most of the RNA in the cell is found in the ribosomes. If most of the DNA is transcribed into ribosomal RNA, one would expect the base ratios of the DNA and the ribosomal RNA in a given species to be similar. They were not. Also the ribosomes seemed to be very similar in different species, but the base ratios of the DNA differed widely from species to species. This anomalous data, confirmed by two groups, challenged the role proposed for the ribosome as a template in the mechanism for protein synthesis. The ribosome did not seem to have the *composition* expected for a template, occupying the intermediate role between DNA and protein in the mechanism.

The compositional anomaly served to localize the problem in the ribosome-as-template stage of the mechanism. However, the compositional anomaly alone did not suggest a redesign hypothesis.

The hypothesis of the messenger RNA resulted from another anomaly for the ribosome as template, a temporal anomaly. This anomaly emerged from work in Paris of Arthur Pardee, Francois Jacob and Jacques Monod (1959) in their famous PaJaMo experiment (discussed in Burian, 1993; Judson, 1996; Morange, 1998). When a functional gene was introduced into the cytoplasm of a strain of bacteria lacking that gene, synthesis of the corresponding gene began very rapidly. Ribosomes are large particles with several subcomponents. There seemed to be insufficient time for the synthesis of a new ribosomal template RNA.

Several hypotheses about the localization of the problem were generated. Each stage in the mechanism prior to the problem became a site to localize the anomaly and suggest revisions (discussed in Olby, 1970). Perhaps in bacteria the DNA itself carried out protein synthesis, without an intermediate template

RNA. Monica Riley (personal communication), the graduate student of Arthur Pardee, who continued this work after he was back at Berkeley, recalls that she thought the DNA was the most likely site for synthesis. Perhaps ribosomes can form more quickly than seemed reasonable. And, finally, perhaps a new type of RNA was rapidly formed, a "tape" or "messenger", as it was called. This RNA would have the base composition of the transcribed DNA, would be formed rapidly, and would play the role of template. Messenger RNA would replace ribosomal RNA in fulfilling the role of template in the mechanism. It occupied a new stage between the DNA and the already formed ribosome.

Again we see the importance in discovery of the generation of alternative hypotheses and the evaluation among the alternatives. Often generation of alternatives during anomaly resolution is easier than at the beginning of the generation process. During anomaly resolution the constraints are tighter; those components not implicated by the anomaly must be retained and the new schema components must be generated to be compatible with them. At some degree of abstraction, it may be possible to generate all possible alternatives that will fit within that constrained context and also function so as to resolve the anomaly.

In the discovery of messenger RNA, there is a famous "a-ha" moment involving Sidney Brenner, Francis Crick and Francois Jacob (discussed in Olby, 1970; Crick, 1988; Jacob, 1988). After the PaJaMo results, the Paris group formed the hypothesis of a hypothetical *X* or tape or messenger RNA. But others were skeptical about the existence of this hypothetical entity as the replacement for the ribosome as template. Jacob was visiting in Brenner's rooms in Cambridge telling Brenner, Crick and others about the experiments. Suddenly Brenner realized that an RNA detected previously by Volkin and Astrachan had the appropriate base composition. When a bacteriophage virus enters the cytoplasm of the bacteria, a new RNA is synthesized that has a base composition like the DNA of the virus, not like that of the host bacteria. Volkin & Astrachan had speculated that the RNA might be a precursor of the phage DNA. But Brenner suddenly realized that the Volkin & Astrachan RNA was the messenger.

The importance of this story for our purposes is that the "a-ha" experience is to be accounted for as the realization that a known entity could play a needed role in a mechanism schema. The role of the template was becoming more and more specified. It should have a base composition like that of the corresponding DNA. The PaJaMo experiment also showed that the template had to be synthesized very quickly. The Volkin & Astrachan RNA satisfied both of these constraints. As happened in other instances of the discovery of the mechanism of protein synthesis, the experimenters who found an entity did not discover its role in the mechanism. (Compare Hoagland's detection of RNA bound to amino acids with Crick's prediction of the adaptor, which is discussed in Hoagland, 1990; Judson, 1996; Rheinberger, 1997; Darden and Craver, 2002.) The "a-ha" moment in this case involved recognizing that an appropriate entity filled a constrained mechanism role. When one is in the throes of anomaly resolution, the problems are very acute. According to Crick's later recollection (Crick, 1988, p. 117), the ribosomal anomaly had been plaguing Brenner and Crick and they had been examining various ways to resolve it. The new constraint from the PaJaMo experiment that the template had to form rapidly added a temporal constraint. This helped to further specify the properties of the template. This new constraint thus aided Brenner in realizing that a previously discovered entity could play the role of template.

Note how the proposed mechanism schema, with its template stage, aids discovery. It indicates the properties that a filler for the role of template must satisfy. It aids in recognizing the anomalies for the ribosome in fulfilling that role. It aids in recognizing that a previously discovered entity could play the role instead.

Sydney Brenner and Francois Jacob planned experiments to disentangle the respective roles of the ribosomes and the messenger in the mechanism of protein synthesis. They carried out the experiments in Mathew Meselson's lab at Cal Tech, using labeling techniques that Meselson had developed (Brenner et al., 1961). The existence of the messenger received support from those experiments, as well as others in Watson's lab at Harvard (Gros et al., 1961).

In their paper of 1961, Jacob and Monod summed up the results of this anomaly resolution episode:

> The property attributed to the structural messenger of being an unstable intermediate is one of the most specific and novel implications of this scheme... This leads to *a new concept of the mechanism of information transfer*, where the protein synthesizing centers (ribosomes) play the *role* of non-specific constituents which can synthesize different proteins, according to specific instructions which they receive from the genes through M-RNA. (Jacob and Monod, 1961, p. 353; emphasis added)

Devising roles and detecting entities that fulfill those compositional and temporal roles go hand in hand in the discovery of all the components of a mechanism.

5. Conclusion

Discovery is to be viewed as a problem-solving process guided by constraints and strategies. Further, discovery is an extended process consisting of cycles of generation, evaluation, and revision. Discovery in biology is often discovery of mechanisms; this paper discussed various constraints and strategies to guide mechanism discovery.

Mechanism sketches and schemata aid the discovery process. First, sketches vividly portray the existence of black boxes, incomplete gaps in the representation of the productive continuity of the mechanism. Thus, they guide the direction of further work to remove such black box incompleteness. Second, schemata provide constraints and guidance in localizing anomalies, and, then further guide anomaly resolution by specifying properties that entities and activities must have to be role fillers in the schema.

Philosophers should move beyond talk of (the lack of) a logic of discovery and a logic of justification to study reasoning strategies for generation, evaluation, and revision in the discovery of mechanisms.

Acknowledgments

This work was supported by the US National Science Foundation under grant SBR9817942 and the General Research Board of the Graduate School of the University of Maryland. Much of my work on mechanisms has been done in collaboration with Peter Machamer and Carl Craver; their ideas pervade this work in ways not fully indicated by citations. I thank Dick Burian, Nancy Hall, Larry Holmes, Hans-Jörg Rheinberger, and Rob Skipper for comments on earlier drafts of this paper; David Didion and Jonathan Roy for research assistance.

References

Bechtel, W. and Richardson, R. C. (1993). *Discovering Complexity: Decomposition and Localization as Strategies in Scientific Research*. Princeton University Press, Princeton, NJ.

Brandon, R. (1985). Grene on mechanism and reductionism: More than just a side issue. In Asquith, P. and Kitcher, P., editors, *PSA* 1984, pages 345–353. Philosophy of Science Association, East Lansing, MI.

Brenner, S., Jacob, F., and Meselson, M. (1961). An unstable intermediate carrying information from genes to ribosomes for protein synthesis. *Nature*, 190:576–581.

Burian, R. M. (1993). Technique, task definition, and the transition from genetics to molecular genetics: Aspects of the work on protein synthesis in the laboratories of J. Monod and P. Zamecnik. *Journal of the History of Biology*, 26:387–407.

Burian, R. M. (1996). Underappreciated pathways toward molecular genetics as illustrated by Jean Brachet's cytochemical embryology. In Sarkar, S., editor, *The Philosophy and History of Molecular Biology: New Perspectives*, pages 67–85. Kluwer, Dordrecht.

Craver, C. F. (2001). Role functions, mechanisms, and hierarchy. *Philosophy of Science*, 68:53–74.

Craver, C. F. (2002). Interlevel experiments, multilevel mechanisms in the neuroscience of memory. *Philosophy of Science* (*Supplement*), 69:83–97.

Craver, C. F. and Darden, L. (2001). Discovering mechanisms in neurobiology: The case of spatial memory. In Machamer, P., Grush, R., and McLaughlin, P., editors, *Theory and Method in the Neurosciences*, pages 112–137. University of Pittsburgh Press, Pittsburgh, PA.

Crick, F. (1959). The present position of the coding problem. *Structure and Function of Genetic Elements: Brookhaven Symposia in Biology*, 12:35–39.

Crick, F. (1988). *What Mad Pursuit: A Personal View of Scientific Discovery*. Basic Books, New York.

Darden, L. (1991). *Theory Change in Science: Strategies from Mendelian Genetics*. Oxford University Press, New York.

Darden, L. (1992). Strategies for anomaly resolution. In Giere, R. N., editor, *Cognitive Models of Science*, pages 251–273. University of Minnesota Press, Minneapolis, MN.

Darden, L. (2002). Strategies for discovering mechanisms: Schema instantiation, modular subassembly, forward/backward chaining. *Philosophy of Science (supplement)*, 69:S354–S365. Preprint available at http://www.philosophy.umd.edu/Faculty/LDarden/.

Darden, L. (2005). Relations among fields: Mendelian, cytological and molecular mechanisms. *Studies in History and Philosophy of Biological and Biomedical Sciences*, 36:357–371.

Darden, L. and Craver, C. F. (2002). Strategies in the interfield discovery of the mechanism of protein synthesis. *Studies in History and Philosophy of Biological and Biomedical Sciences*, 33:1–28.

Glennan, S. S. (1996). Mechanisms and the nature of causation. *Erkenntnis*, 44:49–71.

Gros, F., Hiatt, H., Gilbert, W., Kurland, C. G., Risebrough, R. W., and Watson, J. D. (1961). Unstable ribonucleic acid revealed by pulse labeling of *E. coli*. *Nature*, 190:581–585.

Hoagland, M. B. (1990). *Toward the Habit of Truth*. Norton, New York.

Jacob, F. (1988). *The Statue Within: An Autobiography*. Basic Books, New York. Translated by Franklin Philip.

Jacob, F. and Monod, J. (1961). Genetic regulatory mechanisms in the synthesis of proteins. *Journal of Molecular Biology*, 3:318–356.

Judson, H. F. (1996). *The Eighth Day of Creation: The Makers of the Revolution in Biology*. Cold Spring Harbor Laboratory Press, Cold Spring Harbor, NY. Expanded Edition.

Laudan, L. (1977). *Progress and Its Problems*. University of California Press, Berkeley.

Machamer, P., Darden, L., and Craver, C. F. (2000). Thinking about mechanisms. *Philosophy of Science*, 67:1–25.

Morange, M. (1998). *A History of Molecular Biology*. Harvard University Press, Cambridge, MA.

Morgan, T. H., Sturtevant, A. H., Muller, H. J., and Bridges, C. B. (1915). *The Mechanism of Mendelian Heredity*. Henry Holt and Company, New York.

Olby, R. (1970). Francis Crick, DNA, and the central dogma. In Holton, G., editor, *The Twentieth Century Sciences*, pages 227–280. W. W. Norton, New York.

Pardee, A. B., Jacob, F., and Monod, J. (1959). The genetic control and cytoplasmic expression of 'inducibility' in the synthesis of β-galatosidase. *Journal of Molecular Biology*, 1:165–178.

Rheinberger, H.-J. (1997). *Towards a History of Epistemic Things: Synthesizing Proteins in the Test Tube*. Stanford University Press, Stanford, CA.

Skipper, R. A., Jr. (1999). Selection and the extent of explanatory unification. *Philosophy of Science (Proceedings)*, 66:S196–S209.

Thagard, P. (1999). *How Scientists Explain Disease*. Princeton University Press, Princeton, NJ.

Watson, J. D. (1968). *The Double Helix*. New American Library, New York.

Watson, J. D. and Crick, F. H. C. (1953). A structure for deoxyribose nucleic acid. *Nature*, 171:737–738.

Wimsatt, W. (1972). Complexity and organization. In Schaffner, K. F. and Cohen, R. S., editors, *PSA* 1972, pages 67–86. Reidel, Dordrecht.

ON THE ROLE OF THOUGHT-EXPERIMENTS IN MATHEMATICAL DISCOVERY

Eduard Glas
Delft Institute of Applied Mathematics
Delft University of Technology
E.Glas@ewi.tudelft.nl

Thomas Kuhn has given us a useful account of the role of thought-experiments in empirical science (Kuhn, 1977). As an example he took a thought-experiment of Galileo, which typically was directed at disclosing a conceptual problem. In contrast to a real experiment, a thought-experiment involves no essentially new empirical information but typically relies on information which is already at hand but not assimilated by the traditional mode of dealing with the world. The application of a concept to a thought-experimental situation that lies outside its paradigmatic range and context may then reveal its failure to fit the full structure of reality. Galileo for instance showed that applying the current (Aristotelean) concept of speed in the habitual way to a certain imagined situation led to confusion and contradiction. 'Thinking through' the thought-experimental situation according to the habitual mode of thinking led to conclusions which failed to fit expectations based on familiar experience. In addition, the thought-experiment suggested particular ways to revise both expectations and theory, a revision which was in fact a mathematization.

Thought-experiments were at the heart of Galileo's method. Even his real experiments make a strong impression of being primarily intended as *demonstrations* that the effects calculated from the theory could actually be produced, whereas these theories had been *discovered* in a quasi-mathematical way, through thought-experimentation. Both Galileo's method and the new science of motion to which it gave rise were molded on the example set by Archimedes, of whom we will come to speak presently.

Thought-experimentation thus establishes a strong link between empirical science and mathematics. It is no coincidence that Imre Lakatos, the great protagonist of 'the radical assimilation of mathematics to science', characterized proofs in informal mathematics as thought-experiments, because they cannot

J. Meheus and T. Nickles (eds.), *Models of Discovery and Creativity*,
DOI 10.1007/978-90-481-3421-2_4, © Springer Science + Business Media B.V. 2009

be 'verified' in the strictly logical sense of the term. But they certainly can be 'falsified' inasmuch as they are liable to criticism and improvement. Mathematics is quasi-empirical in that it involves the construction of arguments 'after the facts', working 'upward' from tentative insights to theories (lemmas, axioms, principles) that warrant or account for them. Mathematics is *like* science, *not* because it is somehow based on sensual experiences, but because it proceeds in a 'hypothetico-deductive' way, through fallible guesses and tests (Lakatos, 1978, p. 65).

Like most of his contemporaries, Lakatos's view of science was centred on theory rather than experiment, and this made him to focus almost exclusively on the *logic* of mathematical discovery. The only reason he gave for calling informal proofs thought-experiments was that they, like real experiments, cannot be verified in the strictly *logical* sense of the term. The explication he gave of the 'quasi-empirical' character of mathematics likewise was phrased entirely in the *logical* terms of the characteristic 'direction' of argumentation. This leads directly to the question I want to discuss here: are proofs in informal mathematics thought-experiments only in this abstract 'logical' sense, or also in a more concrete 'procedural' sense? Are they really, intrinsically, similar to scientific experiments, or is this only a figure of speech to emphasize their provisional, not definitive, character? In any case, the role of thought-experiments in mathematics is not exhausted by delivering the proofs and refutations that Lakatos's logic of mathematical discovery relies upon; some have more to do with clarifying and making sense of conceptual issues and ways of handling mathematical objects than with logical connections between statements.

I shall begin with a brief exposition of a case which undoubtedly represents a genuine thought-experiment in mathematics, a piece of informal reasoning which literally might be construed as idealization of a real, a physical experiment: the 'weighing' of the parabola, through which Archimedes established the area of the segment of a parabola. After having specified what I consider to be its most characteristic features, I will focus on a case which usually is not regarded as a thought-experiment at all but, quite the opposite, as a specimen of purely formal development: the introduction of complex numbers in algebra. Comparison of these two very different cases provides us with a basis for discussing the question in what sense and manner mathematical discovery may be construed as a thought-experimental process.

1. Archimedes's Method

A familiar and very typical example of a thought-experiment in mathematics is provided by Archimedes. In his treatise *On Method*, he described for a colleague-mathematician how he first *found* the area of the segment of a parabola by applying an analogy from statics. Archimedes's reasoning was not a linear chain of deductive arguments but involved a complicated network of relations within and between geometrical objects in a certain configuration. He thought of line-segments parallel to the axis of the parabola and bounded by the segment of the parabola as weights and by mobilizing the said network of relations was able to 'balance' them with corresponding line-segments of a certain triangle defined by the parabola. As 'all' the line-segments making up the figures could thus be set in equilibrium, the segment as a whole could be balanced against the total 'weight' (area) of the said triangle, placed in its centre of gravity. The law of the balance then gave the ratio between the area of the triangle and that of the segment of the parabola. The area of the segment turned out to be $4/3$ of an inscribed triangle (T) with the same base and height as the segment (Dijksterhuis, 1987, chapter X).

Archimedes regarded his method as a way of exploring, not of proving. This was not because it involved mechanical notions, but only because the reasoning with 'indivisibles' (taking an area to be made up of line-segments) lacked demonstrative force. Indeed, in his treatise *Quadrature of the Parabola* (Dijksterhuis, op. cit., chapter XI) he demonstrated the same theorem once more by means of statical considerations, but this time without using indivisibles, and here the argument was presented as a geometrical proof satisfying all requirements of exactitude.

Still, the thought-experiment was of vital importance not only for discovering the mathematical 'fact' in question, but *for finding the proof* as well, inasmuch as the outcome entered in the reasoning itself through which the crucial lemma for the proof was constructed. It was a typical instance of constructing an argument 'after the fact', i.e., using the conjectured fact for finding a way to arrive at it. The procedure can be reconstructed as follows:

The inscribed triangle (T) cuts off two new segments, in which triangles with the same base and height as these segments can be inscribed. Their areas taken together can be shown to be $(1/4)T$. The same procedure can be applied to the four new segments thus obtained, etc. Therefore, after n steps the total area covered by the triangles will be $(1+1/4+1/16+\ldots+1/4^n)T$. Now, assuming the outcome of the thought-experiment to be true, the segments generated in every step will exceed their inscribed triangles by 1/3, so that adding $(1/3)(1/4^n)$ to the last term of the series should yield the hypothesized value $(4/3)T$. And indeed, the last two terms of the series $1+1/4+\ldots+1/4^n+(1/3)(1/4^n)$ add up to $(4/3)(1/4^n)$, which is $(1/3)(1/4^{n-1})$, and this added to the previous term in the series gives $(1/3)(1/4^{n-2})$, etc. The sum of the whole series therefore is $1+1/3=4/3$. Accordingly, the area covered by the triangles after n steps is $4/3-(1/3)(1/4^n)$ times T. To this lemma, found by using the outcome of the thought-experiment as guiding hypothesis, the double argument 'ex absurdo' could be applied that finally delivered the exact proof.

This kind of proof required the adoption of a lemma of Euclid's: 'if from a quantity is subtracted more than its half, from the remainder again more than its half, and so on, it will at length become smaller than any assumed quantity'. The proof then consisted of showing that the area of the segment cannot possibly be smaller than $(4/3)T$, for however small one assumes the difference to be, the term $(1/3)(1/4^n)$ can, in virtue of the lemma, always be made smaller still by taking n great enough, and it is impossible that the area covered by the inscribed triangles would be greater than the area of the segment. The assumption that the area of the segment is greater than $(4/3)T$ is refuted in a similar fashion. Therefore, the area of the segment cannot possibly be either greater or smaller than $(4/3)T$, hence it must necessarily be $(4/3)T$.

Characteristic of Archimedes's thought-experimental way of proceeding was the melding of different, geometrical and statical, ways of representing things and reasoning about them, and this 'experimental' procedure made it necessary to warrant the results *post hoc*. As we saw, it was the experiment itself which delivered the tools for the construction of a rigorous proof. The thought-experiment was not merely a suggestive aid in discovery, but essential also for the mathematical justification: the crucial lemma for the proof was constructed 'after the fact' which delivered the essential structuring and guiding assumptions for its construction.

Although the prefix 'thought-' might suggest otherwise, thought-experiments need not literally be performed 'in thought' in the sense that they somehow involve mental representations or images. Archimedes 'thought of' line segments and areas as possessing weight, but this is not essentially different from normal geometrical practice, in which lines are 'thought of' as possessing length but not breadth, etc. The way in which he conducted his statical argument in the *Quadrature of the Parabola* was in all respects similar to that of constructing a rigorous geometrical proof. Archimedes also based his statics proper (the theory of the lever) on postulates—not empirical laws—in exactly the same way as Euclid had axiomatized plane geometry.

That imagery, mental or experiential, is not essential, will become apparent from the case to which I now turn and which typically concerns a field dealing with unimaginable things. I will argue that in a number of relevant respects the introduction of what were called 'impossible numbers' in algebra may be construed as a thought-experimental process.

2. Impossible Numbers

Contrary to what is often thought, complex numbers were not introduced merely to provide 'imaginary' solutions to quadratic equations with negative discriminants. They were indispensable for finding even the perfectly *real* solutions to cubic equations. Sixteenth-century Italian mathematical artists had found, apparently by sheer trial and error, a procedure for solving these equations (Van der Waerden, 1985, pp. 52–59). Leaving historical details aside, the procedure begins by eliminating the quadratic term from the general third-degree equation ($y^3 + py^2 + qy + r = 0$) by a simple substitution ($y = x - p/3$). In the reduced equation $x^3 + ax + b = 0$ (in which a and b are expressions in p, q and r of the original equation) the unknown x is then replaced by the sum of two other unknowns, $x = u + v$, yielding: $u^3 + v^3 + (3uv + a)(u + v) + b = 0$.

The use of replacing one unknown by two unknowns is that although their sum is fixed their product is not (up to a certain boundary) and hence may be 'chosen' freely. Choosing $uv = -a/3$ reduces the equation to $u^3 + v^3 = -b$; writing the former expression as $u^3v^3 = (-a/3)^3$, we get two 'simpler' equations in u^3 and v^3.

In applying this procedure, however, one frequently stumbled on 'impossible' square roots of negative numbers. Even if an equation was evidently satisfied by a real number, the procedure might well yield an 'impossible' expression containing cube roots of what is now known as conjugate complex numbers. Take for instance the simple equation $x^3 = 15x + 4$. Applying the recipe yields $u^3 + v^3 = 4$ and $u^3v^3 = 125$, from which we get the quadratic equation $(u^3)^2 - 4(u^3) + 125 = 0$. This equation, however, appears to have no real solution as its discriminant is negative, and yet the cubic equation whose solution depends on it is evidently satisfied by the real number 4. Here we clearly are facing a profound conceptual riddle.

In his *l'Algebra*, the Italian Bombelli (1572) dealt with it in the following way. If we ignore for the moment the conceptual difficulties and in a quasi-formal manner apply the ordinary rules of algebra to the above example, we find as roots of the quadratic equation: $u^3, v^3 = 2 \pm 11\sqrt{-1}$. For x we accordingly find the 'complex' expression

$$x = \sqrt[3]{2 + 11\sqrt{-1}} + \sqrt[3]{2 - 11\sqrt{-1}},$$

which somehow must 'represent' the real number 4.

Although there is no way actually to compute the two cube roots, their symmetry, and the requirement that their sum must be equal to 4, made Bombelli to *guess* that they should both be of the form $2 \pm n\sqrt{-1}$, in which case each term contributes one half of the required outcome and the imaginary parts vanish in the summation.

Raised to the third power, this form yields $(2 \pm n\sqrt{-1})^3 = 8 - 6n^2 \pm (12n - n^3)\sqrt{-1}$, which is equal to $2 \pm 11\sqrt{-1}$ when $8 - 6n^2 = 2$ and $12n - n^3 = 11$. From the first equation we get $n = \pm 1$ and $n = 1$ appears to satisfy also the second. The hypothesis is therefore confirmed, and we find as solution:

$$x = \sqrt[3]{2 + 11\sqrt{-1}} + \sqrt[3]{2 - 11\sqrt{-1}} = 2 + \sqrt{-1} + 2 - \sqrt{-1} = 4.$$

Bombelli's procedure typically consisted in constructing an argument 'after the facts', that is: a piece of reasoning 'backwards' from the required result to the argumentative steps that were necessary to attain it. The validity of his manipulations with 'impossible' numbers was thus tested by whether they actually led to the required outcome. The real root of the equation had to be known beforehand, because otherwise the cube roots could not be worked out at all. Their 'calculation' therefore had the character of a tentative 'explanation' rather than of a 'derivation'.

Bombelli's way of proceeding was science-like in that it was aimed at accounting for, explaining or making sense of known facts in a *post hoc* fashion. Although no propositions and proofs were involved, I still would maintain that Bombelli made an important 'experimental' discovery. When the formula for the solution of a quadratic equation leads to square roots of negative numbers, one may simply conclude that it has no real solution; there is nothing particularly problematic about that. But when the same happens in the case of a cubic equation, the conclusion that it has no real solution is evidently false. Bombelli discovered a profound conceptual mystery and a way to tentatively deal with it, which led to important new insights into the general conditions of solvability of cubic equations and other algebraic insights which could not be formulated without complex numbers.

Methodological indispensability was the main reason also for Descartes and other 17th-century mathematicians to admit 'imaginary quantities' in algebra (the term first appeared in Descartes, 1637, p. 86). Without them, the rule, for instance, that an nth-degree equation cannot have more than n roots, and other principles of outstanding problem-solving and unifying significance, could not be stated (let alone proved) as rules of general validity (Van der Waerden, 1985, chapter 3). Mathematicians of the 18th century established many new relationships of great generalizing and explanatory power, for instance those between exponential and goniometric functions (as expressed in the well-known Euler formulas, which could only be stated in complex form).

The vigorous but strikingly unrigorous reasonings through which Euler and others had discovered these formulas certainly qualify as thought-experiments. Far from being infallible, they typically were attempts to extend, mainly through analogical reasoning, the rules of ordinary algebra beyond their original range of application. The outcomes of these 'experiments' were tested by requirements of consistency with approved 'laws' of the paradigmatic domain, and appraised by heuristic criteria such as their problem-solving, unifying, generalizing and explanatory potential.

3. Conclusion

What I have done is not simply an attempt at characterizing thought-experiments in mathematics by comparing various cases. A full inventory of all varieties and uses of thought-experimentation in mathematics would have required far more space than I am allowed here. Instead, I have focussed on two extremes: on the one hand a piece of reasoning that may be considered an undisputed paradigm case of a genuine thought-experiment, and on the other hand a case which is not ordinarily viewed as a thought-experiment at all but rather as a paradigm case of purely formal development. I have tried to show in what sense and manner even this 'extreme' case might be construed *as* a thought-experiment. It was an attempt to 'stretch' concepts and rules of ordinary algebra beyond their paradigmatic range and context of application. In this case the results were not warranted by a demonstrative proof, but tested by requirements—apart from consistency with approved principles of the standard domain—mainly of problem-solving and generalizing fruitfulness.

Thought-experiments are not merely 'heuristic'; they are not just suggestive aids in discovery, but are also essential for (and sometimes constitutive of) the argumentation through which their outcome is warranted 'ex post facto' (as in the case of Archimedes). The demonstrative argument is constructed 'after the fact' and often derives its force from the experiment itself. Heuristic and justificatory procedures are complementary, the former necessitating the latter, and the latter depending on the former for their crucial structuring and guiding assumptions.

In mathematics and science alike (thought) experiments are attempts *at once* to 'prove' (to test) a theory *and* to 'improve' it. There is no *essential* difference between 'discovery' and 'justification', between 'trying out' new concepts and methods and seeking to 'warrant' the results of these trials. Indeed, what counts as a warrant or mathematically valid proof is far from historically invariant. Justification is not absolute: proofs in informal (contentual) mathematics do not justify us to accept a result unconditionally, but they justify us to accept it provisionally till it is *improved* by a new thought-experiment. The improvement is a 'refutation' of the previous result only in the 'ex post facto' (some would say: Pickwickian) sense that it shows the latter to be lacking in generality and scope, deficient in unifying, explanatory and problem-solving power, etc. from the point of view of the new result, which implies not only an improved proof but a 'better' theorem as well.

Mathematics proceeds through 'trying and testing', indeed, but in a somewhat wider, less theory-centred sense than Lakatos envisioned. Not all scientific experiments are tests of theories, and the same is true of thought-experiments: not all informal trials and tests in mathematics are 'imaginative test-thought-experiments creating the tools for a proof-thought-experiment' (Lakatos, op.

cit., p. 96). Apart from deductive connections between propositions and axioms or lemmas, thought-experiments typically may involve also far more complicated networks of functional interdependencies. Mathematics grows not only in breadth, by expanding the body of propositions so as to capture greater and greater ranges of problems and questions. Discovering new propositions and proofs is but one of the ways in which mathematical knowledge may grow, though certainly one of fundamental importance. But mathematics grows also in depth, by drawing on resources from different scenes of inquiry and melding them so as to tackle problem situations that fall across the boundaries between them. Archimedes brought mechanical concepts and modes of reasoning to bear on problems of pure geometry. Bombelli extended the standard algebraical concepts and modes of reasoning beyond their paradigmatic domain of countable and measurable objects. In both cases there was a tentative melding of different scenes of inquiry that delivered the major analytical tools for progress towards more comprehensive, integrated and unified theories, enabling more 'profound' understanding (hence growth 'in depth').

References

Bombelli, R. (1572). *L'algebra*. Feltrinelli, Milan. Reprinted 1966.

Descartes, R. (1637). *The Geometry of René Descartes, with a facsimile of the first edition*. Dover, New York. Reprinted 1954, translated D. E. Smith & M. L. Latham.

Dijksterhuis, E. (1987). *Archimedes*. Princeton University Press, Princeton.

Kuhn, T. (1977). *The Essential Tension: Selected studies in scientific tradition and change*. University of Chicago Press, Chicago, London.

Lakatos, I. (1978). *Mathematics, Science and Epistemology*. Cambridge University Press, Cambridge. Edited by J. Warrall and G. Currie.

Van der Waerden, B. (1985). *A History of Algebra: From Al-Khwarizmi to Emmy Noether*. Springer, Berlin.

EXPERIMENTAL SYSTEMS, INVESTIGATIVE PATHWAYS, AND THE NATURE OF DISCOVERY

Frederic L. Holmes

There have been many calls in recent years, from historians, philosophers, and sociologists of science, to be more attentive to the role of experimentation in the development of science. Allan Franklin's complaint in 1986 about the "general neglect of experiment and the dominance of theory in the literature on the history and philosophy of science" (Franklin, 1986, p. 1) has been repeatedly cited and echoed (Le Grand, 1990, p. ix). Although earlier historical studies of experimental science are themselves neglected in some of these calls for a new start, there seems little doubt that in recent years there has been a shift in balance between an earlier emphasis on scientific thought and a current interest in scientific practice which highlights experimental practice.

Often those who reexamine experimentation in a historical setting focus on a single experiment, or a "crucial experiment" supported by a small set of "subsidiary" experiments. Certain experiments, such as the Michelson-Morley, or the Millikan oil drop experiment have achieved historical recognition approaching that of major theories. In 1981 the philosopher Rom Harré published a book entitled *Great Scientific Experiments: 20 Experiments that Changed our View of the World*. Harré was careful to point out that experiments are not "isolated events", but "steps in a sequence of studies through which a delineated subject matter is explored" (Harré, 1981, p. 12). Nevertheless, in each case one experiment stands out as the climactic event, for which the preceding steps serve mainly as preparation.

This predisposition to pick out single "great" or "crucial" experiments is, I believe, linked to the view that the primary role of experiments is to test theories. In his well-known discussion of "Theory and Experiment" in *The Logic of Scientific Discovery*, Karl Popper wrote:

J. Meheus and T. Nickles (eds.), *Models of Discovery and Creativity*,
DOI 10.1007/978-90-481-3421-2_5,

> The theoretician puts certain definite questions to the experimenter, and the latter, by his experiments, tries to elicit a decisive answer to these questions, and to no others. All other questions he tries to exclude [...] Thus he makes his test with respect to this one question as sensitive as possible, but as insensitive as possible with respect to all other associated questions. (Popper, 1961, p. 107)

That Popper used the word "experiments" in the plural here seems to suggest little more than that it may take several tries before the experimenter achieves the optimal conditions stipulated to perform the crucial experiment. The implication that the experiment, or set of experiments in question constitutes a closely bounded event rather than a nodal point in an ongoing, open-ended experimental inquiry, is hard to miss.

The view that experiments are designed to test theories has appeared to be supported by one of the most famous early models of experimental science, Isaac Newton's *Optics*. Each of the experiments described in that work was designed to prove one of a carefully structured series of propositions. Alan Shapiro has shown, however, that Newton was not recounting singular observations or experiments, but recasting his results in a style of exposition customary in the mixed mathematical sciences. From his surviving notebooks one can tell that he reconstructed his discoveries to make them appear, in this case as illustrative of a formal deductive method, in other cases as straightforward Baconian inductions (Shapiro, 1996).

In 1983 Ian Hacking mounted what has since been seen as a timely and effective challenge to the Popperian view of the relation of experiment to theory. Despite the fact that experiment was declared in the seventeenth century to be the "royal road to knowledge", Hacking wrote: " History of the natural sciences is now almost always written as a history of theory. Philosophy of science has so much become philosophy of theory that the very existence of pre-theoretical observations or experiments has been denied." Declaring that he hoped to lead a movement that would "attend more seriously to experimental science", Hacking proclaimed that "Experimentation has a life of its own" (Hacking, 1983, pp. 149–150). He made "no claim that experimental work could exist independently of theory", but that there are many relations between theory and experiment; "some theory precedes some experiment, some experiment and some observation precedes theory, and may for long have a life of its own" (Hacking, 1983, pp. 158–160).

I want to put more emphasis than Hacking does, either in these general statements or in the examples he discusses, on the *long* life of experimentation. If experimental ventures have such longevity, if they cannot be framed within the bounded context of tests of a given theory; cannot be restricted to the attempt to give a decisive answer to a single question, then we cannot in general understand experimentation historically by fixing on single experiments or short sequences of experiments.

In their influential study of experimentation, *Leviathan and the Air Pump*, Steve Shapin and Simon Schaffer treat the experiments of Robert Boyle with that instrument, not as tests of theory in the Popperian sense, but as the "experimental production of matters of fact" (Shapin and Schaffer, 1985, p. 23). But they nevertheless seek to characterize Boyle's experimentation by analyzing only two of the more than forty experiments Boyle recorded in the first volumes of his *New Experiments Physico-Mechanical Touching the Spring of the Air*. If one reads through Boyle's volume, however, it becomes clear that the shape of any one or two of his experiments cannot be understood in isolation from the rest. The experimental venture did take on a "life of its own", in which the relation between theory and experiment was continuously varying. Whether it was Boyle's general theory of the spring of the air, or one of the many subsidiary hypotheses he framed to explain some curious phenomenon observed, sometimes the idea led the experiment, sometimes the experiment spawned a new idea. Some experiments were sharply focused to provide decisive answers to well-posed questions, but some were more loosely exploratory, trying various things to see if something unexpected would turn up (Boyle, 1662).

That experimental investigations are more often long-lived ventures than tightly bounded tests of theories is partly intrinsic to the quest for deep knowledge of the natural world, but partly also a consequence of how scientific careers are organized. Beginning with the structure of the Academy of Sciences in Paris in the 1660's there came into being, at first a very small group, of scientists who were supported with instruments, workplaces, and salaries, and expected to make sustained contributions to science. The first generation Academicians attempted to conduct collective projects in which individual contributions remained anonymous. After the reorganization of 1692, individual members were expected to report to the group regularly on the progress of the personal projects they had taken up. I would argue that this social structure was a strong incentive to transform occasional experiments into continuous research. With the professionalization of science during the nineteenth century this became the dominant pattern for the pursuit of experimental science.

The assertion that experimentation has a long life of its own has profound implications for our effort to produce historical accounts of the experimental sciences. If it is correct, then we can no more understand a single experiment, considered in isolation from an extended series of prior experiments, than we can understand a day or a year in the life of a person without reference to her previous development.

In the history of physics, besides Allan Franklin himself, Peter Galison has elucidated the "scope of experimental autonomy", telling a "story [...] about experimental life that can capture laboratory concerns that have little to do with high theory" (Galison, 1987, pp. 6, 12). In the history of biology, the most sustained effort to develop a framework for the historical interpretation of experimental life has been that of Hans-Jörg Rheinberger. Because I have also worked predominantly in the history of the life sciences, I will concentrate my attention on Rheinberger's position, primarily as worked out in his book *Toward a History of Epistemic Things*.

The central organizing principle of Rheinberger's approach is the idea that "experimental systems" are the "smallest integral units of research". Paraphrasing Hacking's statement, Rheinberger asserts that once a scientist has chosen an experimental system, that system takes on a "life of its own", as the unfolding of its capacities and limitations increasingly define what the scientist can and cannot do, and often lead the investigator in directions that she could not have foreseen in advance. Rheinberger has not invented this unit, because laboratory scientists, particularly in twentieth century biology, routinely describe their work in terms of their experimental systems (Rheinberger, 1997, pp. 25–28). To elucidate the role of the experimental system, however, Rheinberger has adopted a bifurcated plan, alternating between chapters devoted to a case history that illustrates the emergence and life of a particular experimental system, and chapters that draw heavily on continental philosophy. I will not follow him through the depths of his philosophical tour, nor the narrative of the case history, but summarize briefly the generalizations he integrates from these disparate contexts.

"Biological research in particular begins", Rheinberger says, "with the choice of a system rather than with the choice of a theoretical framework" (Rheinberger, 1997, p. 25). The system is not simply there, but must be devised. It is not only an arrangement of instruments, but a system of manipulation that includes conceptual as well as material elements. An experimental system must be sufficiently controlled to reproduce existing phenomena reliably, but not so completely controlled that it is a mere technical device. It must also produce differential results that generate new phenomena, that make the "future":

> an experimental arrangement must be sufficiently open to generate unprecedented events by incorporating new techniques, instruments, model compounds, and semiotic devices. At the same time it must be sufficiently closed to prevent a breakdown of its reproductive coherence. It has to be kept at the borderline of its breakdown. (Rheinberger, 1997, p. 80)

The experimental system takes on a life of its own which leads the investigators who use it to unexpected conjunctures, bifurcations and other events that shift their course in new and unforeseen directions.

No sooner has an outcome been reached, however, than investigators mentally reorganize what has gone before to make the result appear as a goal reached logically through systematic procedures. Historians, too, find it difficult to recover the openness, the groping quality of experimentation that appears afterward aimed consistently at what one previously could not foresee. Much of Rheinberger's subtle analysis, his resort to ideas borrowed from Derrida, Heidegger, and others, is intended to find ways to free historical descriptions of the experimental life from retrospectively imposed closure.

The other novel organizing idea in Rheinberger's analysis is the "epistemic thing", adapted from George Kubler's *The Shape of Time: Remarks on the History of Things*, which focuses on the objects of art rather than their creators. Rheinberger defines epistemic things as the objects of scientific research. They are partly material, partly conceptual, and they are continually redefined as experimentation reveals previously unknown properties. Once they become stabilized, they are no longer the objects of scientific research, but technical things (Rheinberger, 1997, p. 31).

The case history to which he applies these conceptions, and on which he also draws to define them, is admirably fitted to his purpose. Between 1947 and 1962, a group of biochemists associated with Paul Zamecnik at Massachusetts General Hospital devised an experimental, cell-free system to study the synthesis of proteins. When they began, they intended to find ways to distinguish protein synthesis in cancer cells from that in normal tissue, but the capabilities of their system soon led them beyond the cancer question to focus on normal protein synthesis. Along the way the system generated an unforeseen object, soluble RNA. As the properties of this object unfolded it became implicated in the process of information transfer from DNA to protein defined by the newly emerging field of molecular biology. During the fifteen year life of their experimental system, Zamecnik and his co-workers thus generated unanticipated scientific objects, and were led from the discipline, language and methods of one field into those of another. They did not know, until they arrived, where their experimentation was taking them.

Rheinberger's historiographical approach deliberately shifts the focus "from scientists to [...] scientific things" (Rheinberger, 1997, 3–4). What is his purpose in doing so? Does he believe that experimental systems literally have a "life" of their own, or is this a metaphor intended only to help us view experimentation from a fresh and unusual perspective? Are experimental systems like the monsters of science fiction, created by scientists but then escaping their control, acquiring goals independent of their putative masters, even in some cases turning on their creators? When he places experimental systems in the foreground, and the experimentalists in the background, is this a framing device, or do the experimental systems really become the primary actors in the story?

During my studies of experimentation, I have framed the activity in a different way. Following what I have sometimes called the "fine structure" of creative scientific activity, I have retraced the daily operations and thought of individual scientists over stretches of time ranging from two to fifteen years. The organizing contour of this activity has become the "investigative pathway", or "research trail". Like "experimental system", these terms come from the ordinary language of science. They are, however, alternative expressions for a single metaphor, likening the progress of the scientist through time to the literal pathway or trail that traverses a spatial territory. I have sometimes likened the reconstruction of an investigative pathway from the information recorded in laboratory notebooks to the reconstruction of the image of a person hiking along a trail from the footprints left on its surface.

Unlike Rheinberger, I have not developed a deep rationale for the organization of experimental activity along the investigative pathway. It is a largely unexamined metaphor for a pattern that has seemed to me to flow naturally from the nature of the documents—primarily notebook records of daily laboratory activity—on which my narratives have been based. By comparing my approach to that of Rheinberger, however, I hope to clarify their respective features and advantages.

There is much that is shared in the two approaches. Both bring scientists and their working tools into intimate partnership. Both concentrate on the interaction between thought and material operations. Both display the inadequacy of histories limited to single "crucial" experiments, both require the historian to follow long sequences of experimentation, both allow for the unexpected, both attempt to exclude retrospective closure from intruding on the open-endedness of science in progress. Unlike Rheinberger, however, I retain the focus on the scientist. Where he calls for a "biography of things, a filiation of objects, [...] as records of the process of their coming into existence" (Rheinberger, 1997, p. 4), I tie the investigative pathway to the biography of the person who follows the trail I attempt to reconstruct.

Both approaches presume a persistent identity, a source of coherence around which to build a story. Where Rheinberger must assert that experimental systems have a life of their own to justify treating them as the integrating units of his story, I can more easily assume that a scientist has a life of her own. But it cannot be assumed in advance that the experimental activity representing a scientific lifetime or an extended portion of it will turn out to be coherent, an integrated progression, or a meaningful subunit of the scientific activity within the field in which the individual scientist has practiced. Just as experimental systems can migrate from laboratory to laboratory, so the leading front in the pursuit of a given scientific problem may migrate from person to person. What one scientist does next might depend more on what another scientist has done last than it does on where the first scientist has previously been.

When I began, in the 1970s, to reconstruct the experiments recorded in the early laboratory notebooks of Claude Bernard, I did not know whether they would fit into coherent progressions, or whether each experiment might seem somehow disconnected from those that preceded it. I was pleasantly surprised to find that, in most cases, the reason for conducting a particular experiment, under the conditions recorded, made sense in terms of the position in which

Bernard had placed himself by what he had done up until then. At the same time, future steps were not fully predictable by past steps. There was room both for persistent aims and unforeseen shifts in direction. The progression that emerged possessed the two essential features linking it to the pathway metaphor—that each step was methodologically and conceptually linked to the previous step, and that the short-term direction changed frequently while still progressing toward a generally outlined horizon whose details were not yet visible.

This experience was repeated when I turned to the laboratory notebooks of Antoine-Laurent Lavoisier and Hans Krebs. Lavoisier set out the general goals for his intended program of research on the processes that fix or release air in February, 1773. During the next 17 years he faithfully pursued those goals. There were many shifts back and forth among them—interruptions in time, retrievals of partial goals dropped earlier. The path was seldom straight, and the distant horizon was barely glimpsed at the beginning, but the steps were those of a continuous pathway in mind and time. Hans Krebs took up a program to apply the manometric tissue slice method to detect the intermediate steps in metabolic pathways in 1930, and followed that quest relentlessly for 50 years. He shifted rapidly and easily from one problem to another within this broadly defined goal. He planned only one day at a time, had no overarching vision of where his investigations would lead him in the long run. Yet his experimental career, looked at from the end of his long life appears as a persistent pursuit of a cluster of related problems, probed ever further. Without a long view of the route ahead, he, nevertheless, traversed what appears in the reconstruction a long, continuous investigative pathway. As in the case of Bernard, almost all of the experiments recorded in the laboratory notebooks of Lavoisier and of Krebs can be understood as steps directly related to the preceding stretch of the pathway. In the case of Krebs one can frequently identify sudden shifts of direction with the impact of outside influences on his daily decisions, but the next steps are seldom disconnected from what had come before.

The continuity of the individual research pathway can be partially accounted for by the investigator's association with an experimental system. Once an investigator has devised and developed such a system, his skills and experience become so intimately connected with it, that the directions in which its properties lead the experimental activity tend to coincide with the opportunities the scientist perceives for his own further progress. But scientists do abandon experimental systems when they reach the limits of the capacities of the systems, when new systems introduced elsewhere offer more powerful capacities for the next steps in an investigative program, or for other reasons. The coherence and persistence of the individual investigative pathway has, I believe, deeper, psychological foundations. Whenever we identify ourselves with certain goals and pursue them for a time, they tend to become part of our own identities, and we tend to continue along the pathways begun, in order to remain ourselves.

The strongly identifiable, distinctive investigative pathways I have been able to trace through the laboratory records of Lavoisier, Bernard, and Krebs are probably not representative of the typical trajectories of more ordinary scientists. Were we to follow the activities of any of the many less eminent investigators who make up most of any given research field, we would probably find less clearly demarcated, less coherent, trails. What each of them did at any particular time would probably connect more directly to what others in the field were doing at about the same time than to a long prior personal journey. But that does not make it misleading to reconstruct the research trails of exceptional scientists, for they are the ones who give direction to what the others do.

Must a historian choose either to present experimental activity as directed by the capacities of experimental systems creating futures unforeseen by those who activate them; or to present it instead as directed by the intention of an investigator marching toward a goal, however dimly perceived? Or can these two approaches be combined? Two of the three investigators whose courses I have followed over extended portions of their careers possessed distinctive experimental systems whose capacities provided much of the opportunities they pursued and some of the limits to what they achieved. Lavoisier adapted to his purpose combinations of traditional chemical apparatus and pneumatic vessels modified from those of Stephen Hales. Much of his work on combustion, calcination, reduction, and respiration was made possible by the various versions of this basic experimental system that he devised. Krebs inherited from Otto Warburg a powerful experimental system consisting of a micromanometer used to measure the gaseous exchanges of a slice of surviving tissue. This system, together with an array of subsidiary procedures, was basic to all of the experiments he conducted during the first decade of his independent career, a period which led him to the ornithine cycle of urea synthesis, the Krebs cycle, and many lesser discoveries. In these situations the life of an experimental system appeared nearly to coincide with the investigative pathway of an individual scientist.

When, on the other hand, the trajectories of the experimental system and that of the individual investigator diverge, which one provides the more meaningful thread of continuity on which to build a historical narrative of creative experimental science? I think there is no general answer to that question, but only particular answers in specific historical situations. The case history that Rheinberger has chosen to illustrate the powerful role of experimental systems seems entirely apt for that purpose. In the case of Lavoisier, Krebs, and Bernard, on the other hand, it seems that the powerful personal drives of the investigators themselves dominated their experimental systems. The combinations of apparatus and procedures they mobilized for their endeavours were essential to their success, but they themselves generated, step by step along their pathways the futures toward which they strode. Even if they could never know in advance how things would turn out, they always had a keen sense of where they were heading.

His shift of attention from the scientist to her experimental system seems to me to make it more, rather than less difficult for Rheinberger to explain how investigations "arrive at new results". His view that the system must be kept at the "borderline of its breakdown", to be under less than full control, follows from his view that it is the system itself that generates "unprecedented events" (Rheinberger, 1997, p. 80). Why then do scientists strive for the most precise, most reliable experimental systems attainable? In the cases with which I have had experience, it is not the experimental system on the verge of breakdown which has been most conducive to advance into the unknown, but that whose stability and precision allowed the investigator to exploit it with most confidence and flexibility. In organic chemistry, when Justus Liebig devised in 1830 a combustion apparatus so simple and reliable that any student could achieve accurate results, the rate at which the composition of new compounds could be determined greatly accelerated. No longer concerned with the problem of whether divergent analyses were caused by divergent methods, chemists could now be confident that they represented instead differences in the nature of the substances analyzed. Tighter control did not reduce elementary analysis to a technological procedure, but to a standardized experimental system more effective in creative experimentation.

Similarly, the manometric tissue slice method was a powerful generator of novelty for Hans Krebs, just because of its precision and reliability, of the ability it gave him to study metabolic processes under more tightly controlled conditions than had been possible with previous methods. The capacity of the system to generate the future is less mysterious than Rheinberger makes it appear, if we restore to the creativity of the investigator the initiative in devising the conditions, variations in procedures, incorporation of new materials, and other modifications that allow new phenomena to appear.

Rheinberger has introduced the notion of "unprecedented events" in preference to "the often used notion of 'discovery'", because the latter is "part of a positivistic lexicon" that he has sought to avoid. In his case history the "emergence of a soluble, small RNA molecule in the cell-free protein synthesis system" is an example of an unprecedented event. "It first appeared as a compound that had *not* been looked for. Subsequently it changed the character of the whole system" from a means to represent intermediates in a metabolic pathway to a representation of a genetic information transfer (Rheinberger, 1997, p. 134).

Avoidance of the term discovery is not itself unprecedented. Other historians have also proscribed its use because they believe that the concept of scientific discovery implies the truth of what is claimed to be found. In their well-known book, *Laboratory Life*, Bruno Latour and Steve Woolgar described even facts as "constructed", not discovered, in the laboratory (Latour and Woolgar, 1979, p. 235). Followers of this viewpoint assert that even such "classic discovery stories" as that of penicillin by Alexander Fleming should be treated "not as discovery but as construction". Penicillin was a constructed object, according to Wai Chen, because after Fleming had found an unexpected appearance in a discarded culture plate, he saw only "properties of 'penicillin' that were considered to be relevant and useful within the framework of this laboratory" (Chen, 1992, pp. 245–246, 286–287).

Rheinberger's conception of the "unprecedented event" provides an important corrective to the constructionist argument. With it he emphasizes that the new objects that appear unexpectedly are independent of the anticipations, interests, and intentions of those who observe them. The observer must follow the unanticipated direction in which the event leads him, rather than to construct from it a "fact" that fits the observer's prior interests. But Rheinberger has, I believe, an additional tacit motivation for substituting "unprecedented event" for discovery. His need to do so arises from his shift from the scientist to the scientific thing. If the experimental system generates results previously unobserved and unanticipated, then to say that it "discovers" those results is incongruous, because discovery is also an act of perception. If we shift our attention back to the scientist, however, it seems to me that at some point she must discover what her system has generated. In his example, the emergence of a soluble form of RNA was *both* an unprecedented event, *and* a discovery. Which term to prefer depends on whether we focus on the system or the experience of the investigator.

I also want to avoid the positivistic lexicon sometimes associated with discovery, but I think we do not have to abandon words whose everyday meaning is clear, just because of some prior philosophical or sociological tampering with them. Two examples of the use of the word discover, listed consecutively in the *Oxford English Dictionary*, can show the way (Onions, 1973, p. 563):

> Harvey discovered the circulation of the blood in 1783.
> He discovered that he had made a mistake in 1892.

The first example is the outcome of a complex historical process. Harvey's own act of discovery includes so many observations, experiments, and inferences, that historians have difficulty isolating from his account of the discovery what constitutes preparative work, what led directly to his recognition that the blood circulates, and what represents historically his efforts to confirm what he had recognized. That it is now generally accepted that Harvey discovered the circulation is the outcome of many responses to the publication of his *De Motu Cordis*, including further observations, debate, and a gradually emerging consensus. The second example can, on the other hand, at its simplest, refer to an act of recognition occurring in a moment.

One might, at first sight, differentiate the two examples by calling one of them a scientific discovery, the other a discovery of everyday life; but scientists also discover, almost every day, that they have made mistakes in their work. These mistakes rarely appear in their publications, but if we follow their investigative pathways they frequently show up. There are many other discoveries along the investigative pathway that do not become public claims. A scientist will discover that his experimental system is working, or that by making certain small changes it will work better. A scientist discovers an effect that may or may not be unexpected, and seeks to modify the conditions of the experiment so as to amplify the effect. She may discover subsequently that the effect is not relevant to the problem she is studying, and so decide not to pursue it further. She may discover that her system is suitable for studying a problem that would divert her from her original intention, and may decide either to defer her original plan to follow the new trail, or may put the latter aside. A scientist's working days are full of such small discoveries. By recapturing them through the reconstruction of investigative pathways, we may restore to the term discovery the ordinary meaning that has been obscured by issues swirling around the ultimate status of discovery as a knowledge claim.

There is no sharp break between the kind of everyday discovery illustrated in the second dictionary example, and the discovery of the circulation used in the first example. Some of the discoveries made along the investigative pathway may occupy the investigator for only a moment, or a few days. Others he may pursue for weeks or months, only to conclude eventually that it is either an artefact or not important enough to spend further effort on. Occasionally one of these discoveries will reach a threshold of significance and confidence that will prompt the investigator to write a paper to report it. A small proportion of those reported will be received with interest and tested further by others in the field, and a few of these will be recognized historically as enduring scientific discoveries. In reconstructing discovery events at all of these levels, historians need not be concerned about the ultimate status of the knowledge claims involved. What we tell are stories about the experiences, from everyday to momentous, that happen to the scientists whose lives we seek to portray.

On what scale must we reconstruct investigative pathways if we are to make visible the spectrum of discoveries, from everyday to landmark, that a creative scientist may make during her career? Are the patterns of discovery, like fractiles, repeated at different degrees of resolution, or are larger scale discoveries integrations of smaller scale ones in which the patterns of emergence are fundamentally different? Howard Gruber and others have stressed that creative work takes a very long time, yet the thought processes that underlie it take place very swiftly. The complexity of reconstructing historically all levels and scales of this activity at once is overwhelming. We can, however, identify "strategic sites" that will enable us to explore two or three orders of magnitude at a time. For example, a scientist, working mainly alone, at a formative stage in his career may publish, over a decade, a series of research papers that includes at least one discovery great enough to transform a subfield of a modern science.

If laboratory notebooks have survived, the research can be followed day-by-day. Scientists commonly date each experiment recorded in such notebooks, a practice that provides a robust chronological structure along which the historian can trace the pathway that the scientist once followed. Even when the notebook contains little more than the operations performed and the data gathered, the historian can often reconstruct the thought behind these actions. To do so, we combine what we can extract from the progression of experiments and the manner in which conditions are varied, whatever other clues we can obtain from contemporary correspondence, the rationale stated subsequently in publications, occasional reflective comments that the scientist may write down in the notebook, and, for near contemporary cases, conversations in which we probe the memory of a still-living scientist.

It often happens that what the record most directly reveals is the emergence of a novel object in an experimental system in the sense that Rheinberger defines such an event. If we return to the distinction I earlier drew between an "unprecedented event" and the discovery that the scientist makes when she recognizes that this event has occurred, we find that laboratory records seldom locate explicitly the act of perception or its accompanying feelings. I once hoped to find comments in the margins of notebook records in which the investigator gave written expression to the excitement, surprise, or disappointment that I supposed would verify that I had identified such a moment of discovery. Most of the time I found none, even when, as in the case of Hans Krebs or Matt Meselson, I was able to confirm with him that a particular experiment on a particular page did record the event in question. Does then the act of discovery itself elude the historian's ability to recreate the past?

Sometimes we can fit stories that scientists tell about such moments sufficiently well to the written record to reproduce a reasonable facsimile of such an event, but these cases are exceptional. Most often we can establish only the proximate conditions under which, and a bounded period of time within which, the discovery must have taken place. The rest we can only imagine through our common understanding of what such a human experience is like. That is not a reason to doubt the reality of acts of discovery. It is only the consequence of the fact that the function that laboratory notebooks serve for scientists does not require them to record their subjective responses to the unprecedented events that they encounter by chance or by design.

The linearity of a typical laboratory notebook record lends itself to the historical reconstruction of linear investigative pathways. Each entry corresponds to a footstep along the metaphorical trail. A scientist working alone makes one experimental move at a time, just as we take one step at a time when we make our way on foot through previously unexplored territory. The correspondence also lends itself well to the linear nature of a written narrative. These satisfying correspondences can, however, mislead us if we take them too literally. If the scientist can make only one move at a time, that does not necessarily mean that she has only one move in mind at a time. Logistical problems may instead prevent her from implementing several facets of her research plan simultaneously. In my reconstructions of the pathways of Lavoisier and of Krebs I have repeatedly encountered patterns in the shifts from one subline to another which suggest that temporally sequential experiments represented mentally parallel lines, rather than changes in direction along a single connected series.

Such parallel endeavours may coexist in the intentions of the investigator, not only over the short time intervals that separate individual experiments, but over long periods in which the investigator is preoccupied with part of his plan while other parts remain pertinent to his long-range objectives. Howard Gruber has denoted this typical aspect of creative work "networks of enterprise". Enterprises, according to Gruber,

> rarely come singly. The creative person often differentiates a number of main lines of activity. This has the advantage that when one enterprise grinds to a halt, productive work does not cease. The person has an agenda, some measure of control over the rhythm and sequence with which different enterprises are activated. This control can be used to deal with needs for variety, with obstacles encountered, and with the need to manage relationships among creator, community, and audience. (Gruber, 1989, p. 11)

I have found the image of the network of enterprise to be particularly apt for understanding the scientific work of Lavoisier. The metaphor of the investigative pathway cannot fully capture this aspect of scientific research, but no metaphor ever fully describes that which it seeks to illuminate.

In my own work I have followed investigative pathways, wherever possible, at the daily level revealed by laboratory notebooks. But this is not the only scale on which they should be traced. The amount of detail easily becomes overwhelming, and the length of the trail that one can reconstruct at such an intimate degree of resolution is limited. For Claude Bernard I followed only one of his several lines of inquiry, over a period of five years. I followed Lavoisier for fifteen years, but did not cover all facets of his interlocked research program. For Hans Krebs I followed his entire research pathway for ten years, from the time he entered the laboratory of Otto Warburg in 1926, until he published his discovery of the citric acid cycle in 1937. Because he carried out two sets of experiments nearly every working day, this was a very long trail. The narrative fills two large volumes, and some of my best friends regard it as unreadable. I have obviously approached the limits of the scope of the method.

Investigative pathways that are still longer, or that extend beyond the work of one individual, must be traced on larger time scales. We can no longer track each footstep, but must be content to identify more widely separated landmarks along the way. Sometimes a convenient unit distance may be that between one published research paper and the next. Published papers are not intended to be historical accounts of the discoveries presented in them. When we can check them against laboratory records, we can easily tell that they reconstruct, reorder, and smooth out the actual pathway. They are not concerned to recapitulate events just as they happened, but to provide the best case available for what has already been found. A long series of publications extending over a scientific lifetime is, however, an accurate record of the scientist's journey along a different scale of events. So long as we are aware that the two scales reveal different levels of investigative activity, we will not be led astray. Such studies are, I believe, badly needed.

I have deliberately inserted here another metaphor, that of life as a journey. In the journey through life an individual often follows many pathways. The power of these spatial metaphors for temporal human activity is that they express the fact that what we do next depends on where what we have done before locates us. Scientific investigators locate themselves through what they have done and learned in all their prior investigations. As in any walk of life, scientists sometimes make fresh starts, change fields abruptly, reeducate themselves or otherwise depart from what they have done until then. But such leaps are difficult and uncommon. More often the investigative pathway can be extended to a metaphor suitable to frame the creative scientific life.

In this paper I have compared and contrasted Hans-Jörg Rheinberger's treatment of the experimental system as the integrating unit of experimentation in the biological sciences, and my use of the investigative pathway metaphor to organize the same kind of activity. These two approaches are, I hope it has been clear, not mutually exclusive. They are complementary ways to view the same events. I have been greatly impressed by his penetrating, imaginative analysis, and reflecting on it has stimulated me to examine more closely another approach that has long seemed to me intuitively natural, but that also requires critical scrutiny.

References

Boyle, R. (1662). *New Experiments Physico-Chemical Touching the Spring of the Air, and its Effects*. H. Hall, Oxford.

Chen, W. (1992). The laboratory as business: Sir Almroth Wright's vaccine programme and the construction of penicillin. In Cunningham, A. and Williams, P., editors, *The Laboratory Revolution in Medicine*, pages 245–292. Cambridge University Press, Cambridge.

Franklin, A. (1986). *The Neglect of Experiment*. Cambridge University Press, Cambridge.

Galison, P. (1987). *How Experiments End*. University of Chicago Press, Chicago.

Gruber, H. E. (1989). The evolving systems approach to creative work. In Wallace, D. B. and Gruber, H. E., editors, *Creative People at Work*, pages 3–24. Oxford University Press, New York.

Hacking, I. (1983). *Representing and Intervening*. Cambridge University Press, Cambridge.

Harré, R. (1981). *Great Scientific Experiments; 20 Experiments that Changed our View of the World*. Phaidon, Oxford.

Latour, B. and Woolgar, S. (1986). *Laboratory Life: The Construction of Scientific Facts*. Princeton University Press, Princeton.

Le Grand, H. E., editor (1990). *Experimental Inquiries*. Kluwer, Dordrecht.

Onions, C. T., editor (1973). *The Shorter Oxford English Dictionary*. Clarendon Press, Oxford.

Popper, K. R. (1961). *The Logic of Scientific Discovery*. Science Editions, New York.

Rheinberger, H.-J. (1997). *Toward a History of Epistemic Things: Synthesizing Proteins in the Test Tube*. Stanford University Press, Stanford.

Shapin, S. and Schaffer, S. (1985). *Leviathan and the Air Pump: Hobbes, Boyle, and the Experimental Life*. Princeton University Press, Princeton.

Shapiro, A. E. (1996). The gradual acceptance of Newton's theory of light and color, 1672-1727. *Perspectives on Science*, 4:68–89.

ABDUCTION AS A HEURISTIC CONSTRAINT

Scott A. Kleiner
Department of Philosophy
University of Georgia
skleiner@uga.edu

1. Introduction

Scientific research is the search for the solution to important problems facing a scientific discipline at a given time. This research draws on intellectual and technological resources provided by the scientific community, which include concepts, preferences for problems to be pursued, accepted theories and empirical information, background beliefs, a technology of manipulation and production, and a history of past research endeavors. Novel contributions to any one or more of these resources can produce the discovery of hitherto unsuspected entities, events, and processes and means of gaining and representing knowledge thereof. Two strategies for drawing on these resources are well known among philosophers of science:

The Kuhnian strategy (Kuhn, 1962/1970) is that scientific practice should be guided by complete consensus on the several components of a paradigm which define a single field or research specialty. Dissent is strongly discouraged, if not vehemently suppressed. Consensus allows a community to focus on one or a few problems, and to gain the advantage of division of labor. Significant novelty is not anticipated, and perhaps cannot be within a monolithic conceptual system. Historically important innovation occurs only after the consensus has begun to dissolve as a consequence of unexpected empirical anomaly. Throughout the duration of the reign of a paradigm little or no effort should be directed to generating or exploring novel concepts, theories, or experimental strategies. The impetus for novelty is generated serendipetously when it is a product of an endeavor aimed at something else, as in the pursuit of orthodox research objectives. Resources for normal research are drawn from a closed system that specifies a single world view, a single set of theoretical and empirical heuristics, and a single language for thinking and communicating. The principal cost of this strategy is that the researcher must face serious anomalies without preparation,

J. Meheus and T. Nickles (eds.), *Models of Discovery and Creativity*,
DOI 10.1007/978-90-481-3421-2_6,

and must generate solutions to them only from resources called into question by the anomaly.

Feyerabend (1974) is a well known critic of Kuhn's monism. He advocates the articulation of alternative theories and practices as a means of discovering possible refuting evidence for the received theories:

> … evidence that is relevant for the test of a theory T can often be unearthed only with the help of an incompatible alternative theory T'. Thus, the advice to postpone alternatives until the first refutation has occurred means putting the cart before the horse. In this connection I also advised increasing empirical contents with the help of a *principle of proliferation*: invent and elaborate theories which are inconsistent with the accepted point of view, even if the latter should happen to be highly confirmed and generally accepted. (Feyerabend, 1974, p. 26)

Feyerabendian pluralism is a kind of heuristic that encourages the contemplation and development of alternative research programs even though orthodox programs are prospering at the time. This heuristic has advantages of preparing a research community in advance for responses to developments that could undermine orthodox practice. A fund of alternative practices provides a wide range of resources from which to draw information and technique. In contrast to Kuhn's strategy, these resources are open-ended, and practitioners are encouraged to draw from outside the confines of their discipline as well as from within. One evident cost of this strategy is that it could encourage researchers to spread limited resources of time and attention too thinly, leaving little energy to pursue any avenue of research to the depth required for significant contributions. Also Feyerabend's strategy of pursuing the contrary of all components of a research practice is far too permissive. All beliefs and practices have an indefinite number of alternatives, and some kind of filtering of these is necessary for pluralistic strategies to effectively utilize always limited resources.

N. R. Hanson (1961) argued that abduction, reasoning to an explanation, is important in scientific discovery as a means of prior appraisal. Abduction does presuppose some articulation of a theoretical structure, and so it cannot be a means of generating such structures *ab initio*. On the other hand, if one is trying to solve a problem using a variety of extant theoretical resources beyond and even incompatible with paradigmatic theory, explanatory promise can narrow the scope of investigation in accord with basic scientific objectives. I shall argue that this heuristic pluralism is manifest in the recent history of evolutionary biology, where the original neo-Darwinian consensus has been faced with proposals of a variety of contrary theoretical claims and research strategies. This pluralism not only exists, but it is also a good thing: To maximize opportunities for significant discoveries even a 'mature' science should encourage a diversity of beliefs and practices. It is a strategic mistake for researchers in the study of evolution to repress dissidence in an effort to present themselves as a community without plausible dissent on basic principles and practices (Smocovitis, 1992; Dietrich, 1998). I shall first discuss some philosophical problems with abduction and then proceed to an assessment of the current situation in evolutionary biology.

2. The Problem of Abduction

Abduction itself is in danger of being too permissive. Gremlins in the attic can explain noises, Divine Design acting on species can explain adaptation, subtle fluids can explain heat conduction, yet none of these explanations is a matter for serious scientific consideration today. More stringent conditions need to be placed on abductive reasoning. Perhaps one condition on an explanation is that the explaining causes exist, a restriction that would rule out the three possible causes just mentioned. However the truth of or evidence for these existence claims can hardly be known when one is trying to isolate plausible theories before expensive steps toward producing such evidence are undertaken. Thus if Hanson's prior filter is to work, there must be other criteria for goodness of explanations, the fulfillment of which might qualify a theory for its explanatory potential. To fulfill their heuristic role, criteria for potential explanations must be justifiable as indicators of likely success in the effort to achieve scientific objectives without presupposing what is to be sought. One might argue in Hanson's behalf that the objectives of scientific research are to formulate theories that provide better explanations, as well as possessing greater precision, predictive power and truthlikeness. Such an argument supposes that explanatory promise can be assessed without the kind or degree of evidence that is needed for final acceptance. Determining explanatory power then would be justified as achieving one of several scientific objectives, the fulfilment of which can be undertaken one at a time.

Feyerabend also recommends that scientists abandon the 'consistency principle' which demands that all plausible hypotheses cohere with background theories and ontologies. It appears that the abductive constraint on contrarian novelties begs the question against Feyerabend and that he would include explanatory power among those methodological criteria that foster the *status quo*. In rejecting what he calls the consistency condition he must also reject abduction as a heuristic for filtering alternative theories.

However, a contrarian theory may have virtues other than external coherence, that confer heuristic reliability. We shall look for these shortly. Also, since beliefs in the background of scientific practice are many in number, the demand for or against external coherence can take more than one form. The strong conservative position would be that a proposed explanation cohere with all background belief, and a strong radical position would be to entertain preferentially the theories that are incoherent with all such belief. Feyerabend would take the strong radical position. A more plausible but weaker position that is neither strongly conservative nor strongly radical would be that potential explanations cohere with some accepted background beliefs, the totality of which need not be mutually consistent, much less coherent. These background beliefs may differ from or even be contrary to those serving current research programs in some field. They might be borrowed from a different field that could have important but not generally known implications for problems in the field under consideration. Embryology and physiological genetics provide grounds to doubt the pan-adaptationism and gradualism that are central to orthodox neo-Darwinian practice. At the same time recent theories of developmental genetics promise to enhance the explanatory power of evolutionary theory by providing causal links between molecular and phenotypic evolutionary processes. Thus a departure from tradition could still enjoy the support of enhanced explanatory power because support for the propositions contrary to the *status quo* are drawn from another discipline.

Eliott Sober (1993) has offered what he calls the maximum likelihood principle as a criterion for a good explanation. If there are competing explanatory hypotheses for some happening, that hypothesis according to which the occurrence is most likely should be preferred. Sober uses this principle to demonstrate the explanatory superiority of design over chance for pre-Darwinian explanations of adaptedness. As design theorists have argued for millennia, chance occurrences are very much less likely to produce adapted structures than are actions guided by design. In general, the occurrence of one event from a super-astronomically large number of random possible occurrences is a poor explanation, particularly for events that occur together repeatedly. A scrap yard can contain all of the components for a Boeing 747, but a tornado is very unlikely to assemble one, much less several such planes. In more general form this principle requires that the cause be effective in generating the effect, what Newton called sufficiency for the effect. Sufficiency need not mean making the effect probable or frequent, for likelihood is a conditional probability and the principle is intended to be applied comparatively. A cause that predisposes to the production of a rare occurrence that might not otherwise occur is sufficient even though the cause, when present, generates the effect with low frequency.

Sober suggests that a second principle is needed to exclude the actions of gremlins and Designing Deities, a principle demanding prior plausibility or probability: An explanatory strategy is supported if it appeals to causes or events that actually exist or are relatively high in their probability of occurrence. If the first principle requires causal regularities or causal relevance to an event, this condition requires appropriate initial and boundary conditions and the actual occurrence of the relation that brings about the effect. Taken at face value the principle recommends, for example, preference for massive volcanic eruptions as more probable causes of the KT boundary than would be an asteroid impact if there were a record of frequent such eruptions, whereas asteroid impacts are believed to be much more rare. However, rarity of causes is not necessarily an explanatory vice: On behalf of the impact theory it might be argued that mass extinctions themselves are relatively rare. The prior plausibility principle actually imposes two requirements: Causes must be actual, that is there must be good reason to believe that they occur. Secondly, their frequency or the time at which they occur must correlate with the frequency or the time of the effect. The actuality of an asteroid impact at the end of the Cretaceous is evident from the presence of geological evidence for a crater off the coast of Yucatan dating to this time. Darwin argued for the actuality of natural selection by appealing to Malthus' principle and heritable variation. As I have suggested, this condition can be question-begging. However, with at least two conditions for explanatory merit, promise can be shown when one but not the other condition is met. One might see the promise in the impact theory from a coincidence between the time of dinosaur extinction and the dating of the anomalously iridium rich KT boundary. This correlation could be coincidental, but scientific bias against coincidences warranted a search for a causal scenario in which asteroid impacts could extinguish flora and fauna in both terrestrial and marine environments. The reason for developing the causal scenario is to demonstrate the effectiveness of the cause, as is required by the maximum likelihood principle. Also one may know that a certain type of cause is actual or likely to occur over a period of time but not know whether it occurs with the appropriate frequency or on the appropriate occasions to correlate with the effects that it might explain.

Analogies are also important in inferring explanatory promise. There are two poles between which analogies go from less to greater abstraction. A material analogy holds between domains D and D' if and only if the same natural kinds of object appear in each. Two objects are of the same natural kind if they share a number of identical but essential properties. Essential properties of an entity, such as a species, should be understood as those which on the basis of factual knowledge should stand as reliable, though not necessarily infallible markers for the entity, where 'markers' are criteria for identification. Early in his inquiries Darwin recognized that the domains of domestic and wild organisms contain sexually and asexually reproducing organisms manifesting heritable traits and frequent heritable variations. Given like inheritance processes, he reasoned that if selection is effective in producing indefinite divergence in the domestic sphere, it should be equally effective as such in wild animals and plants. This analogy supports Darwin's conclusion that selection is a sufficient cause for speciation, and it is thus an essential step in his abductive argument for natural selection. The analogy indicates that there should be a natural process of speciation comparable to the process of producing domestic varieties. More abstract analogies can hold between domains D and D' without property or substantial identity, for example the 'physical' analogies Clerk Maxwell used in developing electromagnetic field theory (Hesse, 1974; Darden, 1982). The extension of selection theory to groups or cell lineages is based on an analogy in which there is property identity, replication and heritable variation, but no substantial identity. Material and the various more abstract analogies give enough ground for believing that the causal or compositional structure of a well known D could occur in less known D', and thus that explanatory strategies working in D might also work in D'.

Finally, consilience has often been cited as an explanatory virtue. An explanatory scheme in a novel domain may provide consilient explanations of several phenomena in that domain which are otherwise believed to be independent. That is, H is a consilient explanation of several prima facie happenings if and only if they are independent on all known rival hypotheses except H, according to which they are in some way mutually dependent. H might describe their common cause. Hypotheses about atomic composition have long been considered promising as providing the only explanatory link between diverse states of matter, or more recently between various chemical kinds.

In sum, good abductive arguments appeal to explanatory schemata which (1) bear material or more abstract causal analogies to known domains, (2) maximize the likelihood of known phenomena, (3) appeal to actual causes whose frequency correlates with that of the effect and (4) are consilient. Arguments meeting these conditions lend initial plausibility to explanatory schemata for which there may not yet be a large number of empirically justifiable applications. Following Feyerabend, one or more of these conditions might be suspended. However without at least one being met, the abductive filter cannot be heuristically effective because of unconstrained permissiveness. Thus, an explanatory scheme could be judged worth investigating on the grounds that one of these conditions is met, and objectives for further research would be to work out details that would demonstrate that it meets the others.

3. Evolutionary Biology

In his work on developmental biology Scott Gilbert sums up some recent and not so recent happenings in evolutionary theory.

> We are at a remarkable point in our understanding of nature, for a synthesis of developmental genetics with evolutionary biology may transform our appreciation of the mechanisms underlying evolutionary change and animal diversity. Such a synthesis is actually a return to a broader-based evolutionary theory that fragmented at the turn of the past century. ... During the mid-twentieth century, population genetics merged with evolutionary biology to produce the evolutionary genetics of the modern synthesis, while molecular genetics merged with developmental biology to produce developmental genetics. These two vast areas, developmental genetics and evolutionary genetics, are on the verge of a merger that may unite these long-separated strands of biology and may produce a developmental genetic theory capable of explaining macroevolution (Gilbert, 1991).

This last merger I will call the ontogenetic synthesis. One of several problems of explaining macroevolution, the production of species and higher taxa, is raised by proponents of punctuated equilibrium, an interpretation of the fossil record (Eldredge and Gould, 1972). Accordingly, there is paleontological evidence that new species, and sometimes higher taxa, appear suddenly without intermediate links to antecedents. Lineages also show most of their morphological evolution at division or speciation; lineages that do not divide show little morphological change over their duration. These claims conflict with Darwinian uniformitarianism, according to which microevolution, evolution within lineages, occurs by the same mechanism and goes at the same rate as macroevolution, the splitting of lineages into divergent taxa. Macroevolutionary processes are held to be simply composed of a succession of microevolutionary events plus an episode initiating divergence. All evolutionary events consist of gradual changes, that is changes that do not involve dramatic shifts in body plan or the kinds of properties that differentiate species or higher taxa.

The inclusion of developmental processes should enhance the explanatory power of evolutionary theory because, among other things, it fills a causal gap between change in genotype, the unit of heritable variation in neo-Darwinian theory, and change in phenotype, the object upon which selection acts. The importance of this causal gap is enhanced by the neutral theory of molecular evolution, according to which most nucleotide substitutions in DNA have no effect on organismal functioning because they take place in non coding segments and because of the degeneracy of the genetic code. Also there is no apparent correlation between rates of molecular and rates of morphological evolution (Wilson et al., 1974). Embryologists and developmental geneticists have long been aware of great complexity in the genotype-phenotype relation and tools for studying this relation have developed dramatically in the last thirty years. These processes of synthesis are similar to the neo-Darwinian synthesis in which concepts and methods from Mendelian genetics were merged with Darwinian natural history and used to enhance the explanatory power of Darwinian theory by filling a gap in our conception of the transmission of characters between generations.

Advocates of the ontogenetic synthesis argue that the study of evolution should focus on the processes by which phenotypes are generated and altered by various kinds of mutation. This attention should provide a more complete understanding of how the variants that feed natural selection are produced and should explain the potential of an organism for evolutionary change. In order to explain the presence of a trait, one must explain not only how a mutant trait can be maintained or spread in a population, but also how it can be generated from the traits of parent organisms undergoing some sort of mutation. In the ontogenetic synthesis molecular methods for identifying, locating and determining the function of morphogenic substances become part of the empirical base for the study of evolution, thus substantiating Feyerabend's claim that considering alternative theories is a means of introducing new kinds of evidence.

The concept of gene regulation is fundamental to developmental genetics and was introduced in the early 1960's in conjunction with Jacob and Monod's discovery of regulation in the activity of the *lac* operon in the bacterium *E. coli*. A regulatory product, a transcription factor, is produced by a regulatory gene and is capable of stimulating or inhibiting the activity of a target gene. Feedback loops can occur in which a quantity of a product is regulated by its action on the gene that produces it. Gene regulation was extended initially by analogical reasoning to eukaryotic cells in the late 1960's (Britten and Davidson, 1969).

An assembly of regulatory interactions between genes in the developing organism can be viewed as producing a cascade of events that begin with the determinaton of embryonic axes by regulatory products from the mother. This cascade of regulatory events coincides with a succession of stages at the cellular level in which cells of the developing embryo become increasingly specialized. The regulatory interactions between genes are mediated by cellular structures that play a role in the transmission and distribution of regulatory products. Some of these structures themselves are products of prior developmental processes. This succession of production and distribution of regulatory products in an initial structure followed by a more complex structure within which a second generation of regulatory products is distributed is a contemporary conceptualization of what were once called epigenetic processes, processes in which complexity is generated. By contrast, in preformationism complexity is not generated but rather presupposed and implemented. This epigenetic process is conceived as an abstract causal structure constructed from known types of regulatory interactions between genes and from what little is known about genetic regulation in model organisms, such as *Drosophila melanogaster*. The details of this process are far from fully known in the model, much less in other organisms, but this scheme is capable of mirroring at a molecular level Ernst von Baer's recapitulation principles dating from the nineteenth century. Regulatory mutations at the beginning of the developmental process can produce major structural changes in an organism because the early developmental stages set a causal environment within which later stages are directed. For example, these early changes lay down organismal axes and thus can differentiate radiate animals, such as jellyfish, which have distinct dorsal-ventral axes but radial symmetry, from bilaterans, which have distinct anterior-posterior as well as dorsal-ventral axes. Mutations in these earlier stages are likely to disrupt processes that normally would occur later in development, and therefore their effects should be only rarely preserved by selection. On the other hand, mutations have less effect on downstream processes as they occur later in development, that is as they affect the more diverse and specialized branches that terminate in adulthood, and are thus more likely to survive selection. Later stage mutations can include the addition of further developmental stages, so the epigenetic cascade is sometimes enlarged as evolution proceeds. In the pre-Cambrian period when the major body plans in animals first appeared the cascade should have been much less extended than it is in presently existing organisms (Raff, 1992). Thus in this period a mutation introducing the bilaterate plan would be an early stage mutation and thus more likely to be viable than the same mutations in later periods with extended downstream consequences.

Because this epigenetic scheme generates divergent outcomes under different initial conditions it is consilient. It also seems to imply the neo-Darwinian claim that evolution proceeds by small steps, which would be caused by mutations in the terminal branches of the network.

However, the distinction between regulatory and other mutations in developmental genetics provides a mechanism by which even late stage mutations can have what would appear to be a saltatory effect on the phenotype, contrary to neo-Darwinian gradualism (Goldschmidt, 1940; De Beer, 1951; Gould, 1977). The distinction also eases but does not eliminate the constraints against early stage mutations because vital components can be rearranged, deleted or multiplied without being disrupted (Raff, 1996). Regulatory mutations can in one generation alter the timing of development and the relative rates of growth of different components of the body. Homeotic mutations can change the number of segments in arthropods or the appendages in those segments; balancers can be converted to wings thereby rendering a fly similar to a more primitive insect. In these and other processes well known to embryologists changes on one genetic component of the developmental system can have amplified morphological effects. Also there is now good evidence that significant early stage developmental mutations do occur rather frequently, as is evident in diverse types of larval stages in closely related sea urchin species. Still the phylotypic stage, a stage after which the basic body plan for an organismis is highly conserved, can be viewed as a constraint within which differentiation occurs both phylogenetically and ontogenetically (Raff, 1996).

Hence phenotypic macromutations can but need not be a product of mutations in early acting regulatory genes, and they can be generated by known epigenetic mechanisms. Nor need they be the summation of many mutations of small effect. Biologists have sometimes been concerned about the low rate of point mutation. If a small fraction of these are functional, as claimed in the neutral theory, and a smaller fraction are adaptive, as would be implicit in the assumption that all mutations are 'random', then there is a problem of commensurability between the supply of functional mutations and supposed rates of evolution, particularly as needed for episodes of evolution in punctuated equilibrium. Any reduction in the numbers of adaptive mutations needed for an evolutionary episode would have potential explanatory advantages, according to the prior probability principle.

Further recent developments in the theory of gene regulation have raised questions about two more neo-Darwinian assumptions: One is that the nucleic acid gene is the only unit of variation and inheritance. The second assumption challenged is that there is no mechanism by which developmental changes can be transmitted to the gametes, genes or offspring (Jablonka and Lamb, 1995). There is now substantial evidence that gene activity can be regulated by heterochromatization, the manner in which the DNA is bound up in chromosomes, and by methylation of DNA regulatory regions. Patterns of methylation and chromosomal heterochromatin are called epigenetic states because they are due to a superstructure superimposed on the coding units of DNA, they are heritable through a templating process in cell division and they are subject to mutation. As regulatory factors they play a role in determining phenotypic characters.

There also is evidence that changing heterochromatic and methylation states of cells initiate specialized cell lineages in the embryo and are thus responsible for the mitotically heritable differential gene activity that is characteristic of cell specialization in development. Another characteristic of these facultative epigenetic states is that they can be induced and even directed adaptively by the environment. Cells can be caused to specialize in specific directions by their cellular environment, and that specialized state is passed to daughter cells as the tissue grows. In development these determinations must be adaptive to be viable. Thus adaptive induced epimutation is the rule in much of metazoan development.

If an organism reproduces asexually, an induced epigenetic state can be passed on to offspring by the same clonal division that occurs in development. Since asexually reproducing organisms reproduce by mitotic division, material analogy supports hypothesizing a mechanism whereby clonally reproducing organisms can evolve by directed and even adaptive mutation, contrary to neo-Darwinian constraints.

Meiotic cell division usually erases these induced epigenetic states, though this erasure process may itself be an adaptation that protects a lineage against readily induced harmful epimutations. An evolved mechanism that excludes somatic differentiation from gametes is not determined by fundamental physio-chemical laws, and thus could readily admit exceptions. Such exceptions would be expected where frequent heritable adaptations of intermediate duration would be advantageous. Indeed, there is some evidence that epigenetic states pass through meiosis because maternally inherited genes often have different activity states than those inherited paternally. Again, by material analogy, some of these that pass meiosis could be directively induced epigenetic states. Thus combining developmental induction with parental marking gives rise to a pathway for directed or even adaptive mutation in sexually reproducing organisms.

Another barrier to the sexual transmission of epigenetic states is Weismann's rule, according to which epigenetic processes in development cannot be transmitted because germ cell lineages are segregated early in development and do not undergo developmental specialization. However there are many exceptions to Weismann's rule in both plants and animals. Late segregating germ lines can stem from epigenetically specialized antecedents, so without evolved protective mechanisms, such as epigenetic erasure in meiosis, there should be various induced epigenetic states available for transmission through the gametes. Hence in many organisms the pathway by which induced epigenetic mutations can be transmitted to offspring may be unobstructed.

In sum, direct evidence shows that induced adaptive epigenetic states occur regularly in development. Direct evidence shows in other contexts that germ cells often derive from epigenetically differentiated somatic cells. Furthermore, in a third situation epigenetic states are transmitted through meiosis. If these three contexts are extended by material analogy so that they overlap, we assemble a mechanism for a kind of Lamarckian genetic response to environmental challenges to an organism. This combination of analogical extension and assembly is literally a strategy for theory construction, and in this case the product is a mechanism for directed mutation in evolutionary processes. The argument should be taken as showing that the mechanism is more than just possible, and that where it occurs it will be effective in producing specific mutations in response to environmental conditions. Thus it meets one of the several conditions on explanatory promise. Another way of looking at the effectiveness of this mechanism is to note that directed mutations would be more efficient than 'random' mutations in producing adaptive structure, and their responsiveness to the environment may be another avenue of explaining punctuated equilibrium.

On the other hand, evidence for the various steps in this argument does seem tenuous at this time, particularly in the sense that epigenetic transmission through meiosis, as evidenced by parental marking, may be too limited in kind or frequency to be of evolutionary significance. The argument superimposes processes that may separately be infrequent, and without further evidence the superposition can only be even more infrequent. The theory of directed mutation thus constructed yet falls short on the prior probability requirement; as yet there is little evidence that it occurs with sufficient frequency to be important in evolution.

However, defenders of this mechanism argue that the paucity of evidence for meiotic transmission of epigenetic states is due to the lack of resources directed to searching for this kind of process. Evidence for this argument is that the type of germ line segregation is known in only about a third of the known species. Lack of attention to the processes on the supply side of evolution may be a product of the social entrenchment of orthodox neo-Darwinism. It appears that, as Feyerabend has suggested, the only avenue for undermining this entrenchment is to encourage those in control of funding and publication as well as promising students to consider and support the pursuit of qualified alternative theories. I submit that abductive argumentation is one among many other means of encouragement, and it does have a virtue of limiting the number of alternatives that might be considered.

4. Conclusions

Several ramifications of the ontogenetic synthesis are contrary to well known constraining assumptions for neo-Darwinian practice. Among these are gradualism, the uniformity of all modes of evolution, the randomness of mutation and the primacy of selection as the cause of diversity, adaptation and the rate and mode of evolution. Although a case is made that selection is not the only explanatory cause of the major features of evolution and its products, I do not think that the proponents of the ontogenetic synthesis can legitimately deny the importance of Darwinian selection as a pervasive constraint on developmental mutants or as a means of explaining the spread of characteristics in a population. However, they appear to have good reason for denying strong adaptationism, the view that selection is the only force driving evolution. Defenders of the ontogenetic synthesis contend that the developmental process generates not only individuals, but also the various kinds of mutant which are then filtered by selection. The epigenetic apparatus that produces the individual organism is the equal of selection in explaining the occurrence adapted form and diversity; both processes are indispensable in such explanations. Gould and Lewontin (1979) advocated developmental constraints on adaptationism. However reflections on the role of ontogenesis in evolution suggest that development is more than just a constraint; it is an active generator of biological form, whereas selection provides the constraint in the form of death for some developmental mutants at various developmental stages, and reproductive success for others. Developmental theories are far from complete, but the several components of the explanatory scheme in the ontogenetic synthesis are supported to some extent by consilience, maximum likelihood, prior probability, and analogy as well as some empirical evidence. This is enough to establish the explanatory potential of newly constructed processes on the mutational side of evolution. Also the several new mechanisms that have emerged suggest that there may be many more new evolutionary mechanisms to come from molecular biology, among them hybridization and disruption of karyotypic stability and transposition of regulatory elements as mechanisms of macroevolution (McCarthy et al., 1995; McDonald, 1990).

References

Britten, R. J. and Davidson, E. H. (1969). Gene regulation for higher cells: A theory. *Science*, 165:349–357.

Darden, L. (1982). Artificial intelligence and philosophy of science: Reasoning by analogy in theory construction. *PSA 1982*, Philosophy of Science Association, 2:147–165.

De Beer, G. R. (1951). *Embryos and Ancestors*. Clarendon Press, Oxford UK.

Dietrich, M. R. (1998). Paradox and persuasion: Negotiating the place of molecular evolution within evolutionary biology. *Journal of the History of Biology*, 31:85–111.

Eldredge, N. and Gould, S. (1972). Punctuated equilibria: an alternative to phyletic gradualism. In Schopf, T. J. M., editor, *Models in Paleontology*, pages 82–115. Freeman, Cooper, San Francisco CA.

Feyerabend, P. K. (1974). *Against Method*. New Left Books, London UK.

Gilbert, S. F. (1991). *Developmental Biology*. Sinauer Publishing Co., Sunderland MA.

Goldschmidt, R. (1940). *The Material Basis of Evolution*. Yale University Press, New Haven CT.

Gould, S. J. (1977). *Ontogeny and Phylogeny*. Harvard University Press, Cambridge MA.

Gould, S. J. and Lewontin, R. (1979). The spandrels of san marco and the panglossian paradigm: A critique of the adaptationist program. *Proceedings of the Royal Society*, 205:581–598.

Hanson, N. R. (1961). *Patterns of Discovery*. Cambridge University Press, Cambridge.

Hesse, M. B. (1974). *The Structure of Scientific Inference*. Cambridge University Press, Cambridge UK.

Jablonka, E. and Lamb, M. J. (1995). *Epigenetic Inheritance and Evolution: The Lamarckian Dimension*. Oxford University Press, Oxford UK.

Kuhn, T. S. (1962/1970). *The Structure of Scientific Revolutions*. University of Chicago Press, Chicago IL.

McCarthy, E. M., Asmussen, M. A., and Anderson, W. (1995). A theoretical assessment of recombinational speciation. *Heredity*, 74:502–509.

McDonald, J. F. (1990). Macroevolution and retroviral elements. *Bioscience*, 40:183–191.

Raff, R. (1996). *The Shape of Life: Genes, Development and the Evolution of Animal Form*. University of Chicago Press, Chicago IL.

Raff, R. A. (1992). Evolution of developmental decisions and morphogenesis: the view from two camps. *Development 1992 Supplement*, pages 15–22.

Smocovitis, V. B. (1992). Unifying biology: The evolutionary synthesis and evolutionary biology. *J. Hist. Bio.*, 25:1–65.

Sober, E. (1993). *Philosophy of Biology*. Westview Press, Boulder CO.

Wilson, A. C., Maxson, L. R., and Sarich, V. M. (1974). The importance of gene rearrangment in evolution: evidence from studies on rates of chromosomal, protein and anatomical evolution. *Proc. Nat. Acad. Sci. U.S.A.*, 71:3028–3030.

CREATIVE ABDUCTION AND HYPOTHESIS WITHDRAWAL

Lorenzo Magnani
Department of Philosophy and Computational Philosophy Laboratory
University of Pavia
and
Sun Yat-Sen University
Guangzhou (Canton)
lmagnani@unipv.it

Abstract Philosophers of science in the twentieth century have traditionally distinguished between the logic of discovery and the logic of justification. Most have concluded that no logic of discovery exists and, moreover, that a *rational* model of discovery is impossible. In short, scientific discovery is irrational and there is no reasoning to hypotheses. A new abstraction paradigm aimed at unifying the different perspectives and providing some design insights for future ones is proposed here: the aim is to emphasize the significance of *abduction* in order to illustrate the problem solving process and to propose a unified epistemological model of scientific discovery. The model describes the different kinds of abductive reasoning and illustrates its formal models in order to classify and analyze the different roles played by inconsistencies in different reasoning tasks. There has been little research into the weak kinds of negating hypotheses. I will consider a kind of "weak" hypothesis that is hard to negate and the ways for making it easy. In these cases the subject can "rationally" decide to withdraw his hypotheses even in contexts where it is "impossible" to find "explicit" contradictions. In these cases the use of negation as failure is illuminating. I explore whether this kind of negation can be employed to model hypothesis withdrawal in the case of the negation of physical unfalsifiable "conventions".

1. Change in Theoretical Systems

In different theoretical changes we witness different kinds of discovery processes operating. Discovery methods are *data-driven* (generalizations from observation and experiments), *explanation-driven* (abductive), and *coherence-driven* (formed to overwhelm contradictions) (Thagard, 1992). Sometimes there is a mixture of such methods: for example, a hypothesis devoted to over-

J. Meheus and T. Nickles (eds.), *Models of Discovery and Creativity*,
DOI 10.1007/978-90-481-3421-2_7, © Springer Science + Business Media B.V. 2009

come a contradiction is found by abduction. Therefore, contradiction and its reconciliation play an important role in philosophy, in scientific theories and in all kinds of problem-solving. It is the driving force underlying change (thesis, antithesis and synthesis) in the Hegelian dialectic and the main tool for advancing knowledge (conjectures and refutations, Popper, 1963, and proofs and counter-examples, Lakatos, 1976) in the Popperian philosophy of science and mathematics.

Following Quine's line of argument against the distinction between necessary and contingent truths (Quine, 1951), when a contradiction arises, consistency can be restored by rejecting or modifying any assumption which contributes to the derivation of contradiction: no hypothesis is immune from possible alteration. Of course there are epistemological and pragmatic limitations: some hypotheses contribute to the derivation of useful consequences more often than others, and some participate more often in the derivation of contradictions than others. For example it might be useful to abandon, among the hypotheses which lead to contradiction, the one which contributes least to the derivation of useful consequences; if contradictions continue to exist and the assessed utility of the hypotheses changes, it may be necessary to backtrack, reinstate a previously abandoned hypothesis and abandon an alternative instead.

Hence, the derivation of inconsistency contributes to the search for alternative, and possibly new hypotheses: for each assumption which contributes to the derivation of a contradiction there exists at least one alternative new system obtained by abandoning or modifying the assumption.

The classical example of a theoretical system that is opposed by a contradiction is the case in which the report of an empirical observation or experiment contradicts a scientific theory. Whether it is more beneficial to reject the report or the statement of the theory depends on the whole effect on the theoretical system. It is also possible that many alternatives might lead to non-comparable, equally viable, but mutually incompatible, systems.[1]

As Lakatos argues, in a mature theory with a history of useful consequences, it is generally better to reject an anomalous conflicting report than it is to abandon the theory as a whole. The cases in which we have to abandon a whole theory are very rare: a theory may be considered as a complex information system in which there is a collection of cooperating individual statements some of which are useful and more firmly held than others; propositions that belong to the central core of a theory are more firmly held than those which are located closer to the border, where instead rival hypotheses may coexist as mutually

[1]Thagard proposes a very interesting computational account of scientific controversies in terms of so-called *explanatory coherence* (Thagard, 1992), which improves on Lakatos' classic one (Lakatos, 1970, 1971); see also Subsection 8, below.

incompatible alternatives. Accumulating reports of empirical observations can help in deciding in favor of one alternative over another.

We have to remember that even without restoring consistency, an inconsistent system can still produce useful information. Of course from the point of view of classical logic we are compelled to derive any conclusion from inconsistent premises, but in practice efficient proof procedures infer only "relevant" conclusions with varying degrees of accessibility, as stated by the criteria of non-classical *relevant entailment* (Anderson and Belnap, 1975).

We may conclude by asserting that contradiction, far from damaging a system, helps to indicate regions in which it can be changed (and improved). It is always better to produce mistakes and then correct them than to make no progress at all. Contradiction has a preference for strong hypotheses which are more easily falsified than weak ones; and moreover, hard hypotheses may more easily weakened than weak ones, which prove difficult subsequently to strengthen. In Section 4 we will consider a kind of "weak" hypothesis that is hard to negate; we will also illustrate the ways for making it easy, by explaining the logical and computational notion of *negation as failure*. In the following two sections we will briefly review abductive reasoning, its formal models, and various ways of governing inconsistencies.

2. Abduction: Sentential, Model-Based, Manipulative

What is abduction? Many reasoning conclusions that do not proceed in a deductive manner are *abductive*. For instance, if we see a broken horizontal glass on the floor we might explain this fact by postulating the effect of wind shortly before: this is certainly not a deductive consequence of the glass being broken (a cat may well have been responsible for it). Hence, *theoretical* abduction (Magnani, 2001) (cf. Figure 1) is the process of *inferring* certain facts and/or laws and hypotheses that render some sentences plausible, that *explain* or *discover* some (eventually new) phenomenon or observation; it is the process of reasoning in which explanatory hypotheses are formed and evaluated.

There are two main epistemological meanings of the word abduction: 1) abduction that only generates "plausible" hypotheses (*selective* or *creative*) and 2) abduction considered as *inference to the best explanation*, which also evaluates hypotheses (cf. Figure 2). To illustrate from the field of medical knowledge, the discovery of a new disease and the manifestations it causes can be considered as the result of a creative abductive inference. Therefore, creative abduction deals with the whole field of the growth of scientific knowledge. This is irrelevant in medical diagnosis where instead the task is to select from an encyclopedia of pre-stored diagnostic entities. We can call both inferences ampliative, selective and creative, because in both cases the reasoning involved amplifies, or goes beyond, the information incorporated in the premises. All we can expect

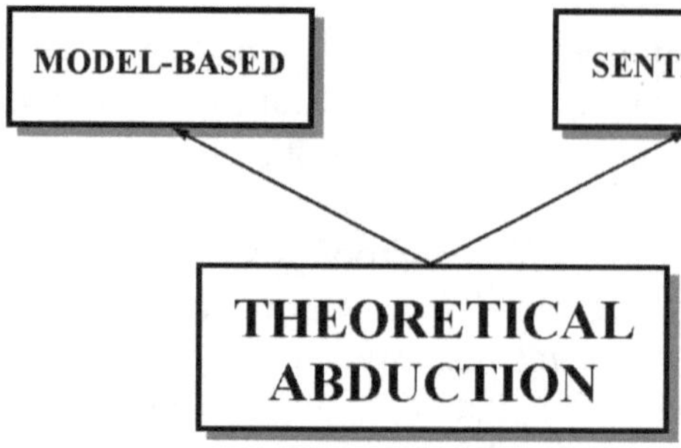

Figure 1. Theoretical abduction

of our "selective" abduction, is that it tends to produce hypotheses for further examination that have some chance of turning out to be the best explanation. Selective abduction will always produce hypotheses that give at least a partial explanation and therefore have a small amount of initial plausibility.

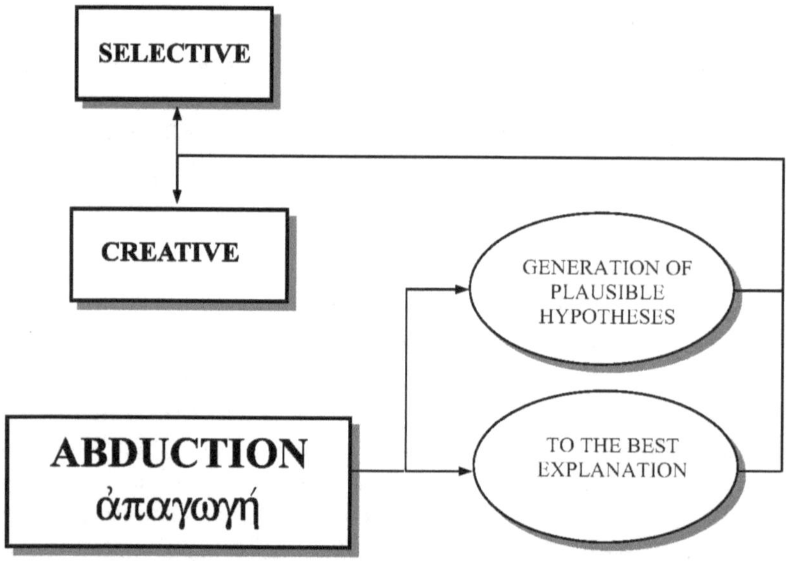

Figure 2. Creative and selective abduction

Finally, many attempts have been made to model abduction by developing some formal tools in order to illustrate its computational properties and the relationships with the different forms of deductive reasoning (see below Section 3.1). This kind of sentential frameworks exclusively deals with selective abduction (diagnostic reasoning) and relates to the idea of preserving consistency. If we want to provide a suitable framework for analyzing the interesting cases of creative reasoning (in science too), we do not have to limit ourselves to the sentential view of theoretical abduction but we have to consider a broader inferential one which encompasses both sentential and what I call model-based elements of creative abduction.

First, it is necessary to show the connections between abduction, induction, and deduction and to stress the significance of abduction to illustrate the problem solving process. I and others (Lanzola et al., 1990; Ramoni et al., 1992) have developed an epistemological model of medical reasoning, called the *Select and Test Model* (ST-MODEL, see Magnani, 1992; Stefanelli and Ramoni, 1992) which can be described in terms of the classical notions of abduction, deduction and induction: it describes the different roles played by such basic inference types in developing various kinds of medical reasoning (diagnosis, therapy planning, monitoring). Abduction is becoming an increasingly popular term in artificial intelligence (Peng and Reggia, 1987a, 1987b; Pople, 1973; Reggia et al., 1983; Thagard, 1988, 1992) especially in the field of medical knowledge-based systems (Josephson et al., 1986; Josephson and Josephson, 1994; Magnani, 1988, 1992a; Ramoni et al., 1992). In the nineteenth century, Peirce (1958) interpreted abduction essentially as an inferential *creative process* of generating a new hypothesis and developed the kind of sentential (syllogistic) model of abduction in terms of abduction, deduction, and induction I will describe below in this section. In the view concerning abduction as inference to the best explanation advocated by Peirce one might require that the finally chosen explanation be the *most* plausible.

Induction in its widest sense is an ampliative process of the generalization of knowledge. Peirce distinguished three types of induction and the first was further divided into three sub-types. A common feature is the ability to compare individual statements: using induction it is possible to synthesize individual statements into general laws (types I and II), but it is also possible to confirm or discount hypotheses (type III). Clearly I am referring here to the latter type of induction, that in my model is used as the process of reducing the uncertainty of established hypotheses by comparing their consequences with observed facts.

Deduction is an inference that refers to a logical implication. Deduction may be distinguished from abduction and induction on the grounds that only in deduction the truth of inference is guaranteed by the truth of the premises on which it is based. All these distinctions need to be exemplified. To describe how the three inferences operate, it is useful to start with a very simple syllogistic (sentential) example dealing with diagnostic reasoning:

1 If a patient is affected by a beta-thalassemia, his/her level of hemoglobin A2 is increased.

2 John is affected by a beta-thalassemia.

3 John's level of hemoglobin A2 is increased.

By deduction we can infer (3) from (1) and (2); by induction we can go from a finite set of facts, like (2) and (3), to a universally quantified generalization, like the piece of hematologic knowledge represented by (1). Starting from knowing—selecting—(1) and observing (3) we can infer (2) by performing a selective abduction. Such an inference is not affected by uncertainty, since the manifestation (3) is pathognomonic for beta-thalassemia. This is a special case, where there is no abduction because there is no "selection". In general clinicians very often have to deal with manifestations which can be explained by different diagnostic hypotheses. The abductive inference rule corresponds to the well-known fallacy called affirming the consequent

$$\frac{\begin{array}{c}\varphi \rightarrow \psi \\ \psi\end{array}}{\varphi}$$

Thus, *selective abduction* is the making of a preliminary guess that introduces a set of plausible diagnostic hypotheses, followed by deduction to explore their consequences, and by induction to test them with available patient data, 1) to increase the likelihood of a hypothesis by noting evidence explained by that one, rather than by competing hypotheses, or 2) to refute all but one (cf. Figure 3).

If during this first cycle new information emerges, hypotheses not previously considered can be suggested and a new cycle takes place: in this case the *non-monotonic* character of abductive reasoning is clear. As stated above, there are two main epistemological meanings of the word abduction (Thagard, 1992): 1) abduction that only generates plausible hypotheses (*selective* or *creative*)—and this is the meaning of abduction accepted in my epistemological model—and 2) abduction considered as *inference to the best explanation*, that also evaluates hypotheses. In the latter sense the classical meaning of abduction as *inference to the best explanation* (for instance in medicine, to the best diagnosis) is described in my epistemological model by the complete abduction–deduction-induction cycle. All we can expect of my "selective" abduction, is that it tends to produce hypotheses that have some chance of turning out to be the best explanation. Selective abduction will always produce hypotheses that give at least a partial explanation and therefore have a small amount of initial plausibility. In this respect abduction is more efficacious than the blind generation of hypotheses.

My epistemological model should be regarded as a very simple and schematic illustration of scientific theory change. In this case selective abduction is re-

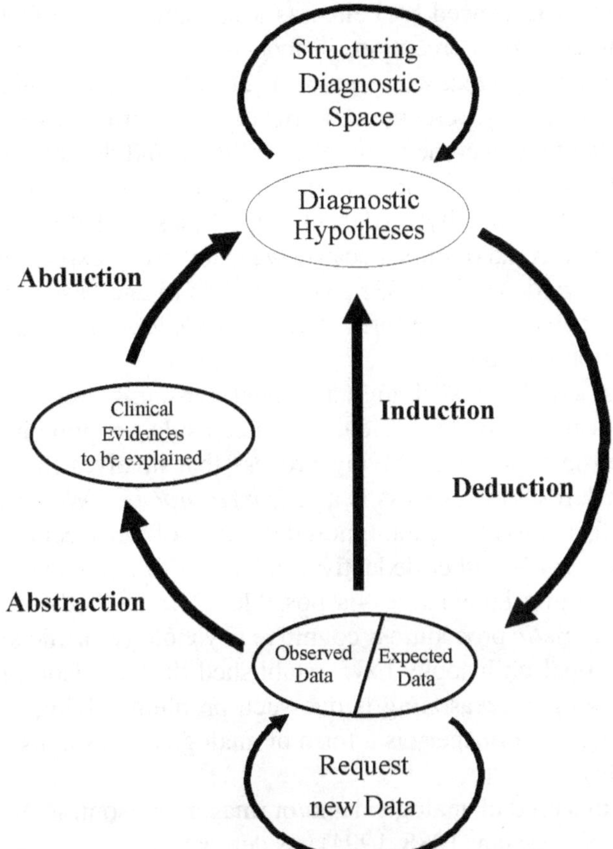

Figure 3. An abductive model of diagnostic reasoning

placed by creative abduction and there exists a set of competing theories instead of diagnostic hypotheses. Furthermore the language of background scientific knowledge should be regarded as open: in the case of competing theories, as they are studied using the epistemology of theory change, we cannot—contrary to Popper's viewpoint (Popper, 1963)—reject a theory simply because it fails occasionally. If for example such a theory is simpler and explains more significant data than its competitors, then it can be accepted as the best explanation (see below, Section 8).

We should remember, as Peirce noted, that abduction plays a role even in relatively simple *visual phenomena*. *Visual abduction*, a special form of abduction, occurs when hypotheses are instantly derived from a stored series of previous similar experiences. It covers a mental procedure that tapers into a non-inferential one, and falls into the category called "perception". Philosophically, *perception* is viewed by Peirce as a fast and uncontrolled knowledge-production procedure. Perception, in fact, is a vehicle for the instantaneous retrieval of knowledge that was previously structured in our mind through inferential processes. By perception, knowledge constructions are so instantly reorganized that they become habitual and diffuse and do not need any further testing. Many visual stimuli are ambiguous, yet people are adept at imposing order on them: "We readily form such hypotheses as that an obscurely seen face belongs to a friend of ours, because we can thereby explain what has been observed" (Thagard, 1988, p. 53). This kind of image-based hypothesis formation can be considered as a form of *visual abduction* (Magnani et al., 1994) (see also Section 8 below).

We have to say that visual and analogical reasoning are productive in scientific concept formation too; scientific concepts do not pop out of heads, but are elaborated in a problem solving process that involves the application of various procedures: this process is a *reasoned process*. We know that scientific concept formation has been ignored because of the accepted view that no "logic of discovery"—either deductive, inductive, or abductive algorithms for generating scientific knowledge—is possible. The methods of discovery involve use of *heuristic* procedures: cognitive psychology, artificial intelligence, and computational philosophy have established that heuristic procedures are reasoned. Analogical reasoning is one such problem solving procedure, and some reasoning from imagery is a form of analogical reasoning (Holyoak and Thagard, 1996).

How does this kind of analogical and/or imagery reasoning function in problem solving? Nersessian (1988, 1994) has demonstrated that history of science abounds with instances of the use of imagery and of analogy to transform vague notions into scientifically viable conceptualizations of a domain. Her analysis deals with the important case of the use of imagery and analogy by Faraday and Maxwell in the construction of the concept of field. The concept of field had its origins in vague speculations about processes in the regions surrounding bodies and charges that might contribute to their action upon one another. In articulating a field representation for electric and magnetic actions, Faraday used primarily qualitative concepts and reasoned from imagery figures. He created a field representation for electric and magnetic actions by reasoning from an imagery representation of the "lines of force" that are formed when iron filings are sprinkled around a magnetic source. Many features of "lines" are incorporated into his field concept. He discussed many actions as "expanding", "bending", "being cut". All the forces of nature are unified and interconvertible through various motions of the lines of forces, and matter, itself, is nothing but point centers of converging lines of force. This representation enabled Faraday to express a quantitative relationship between the number of lines cut and the intensity of the induced force. At the end of this research Faraday intro-

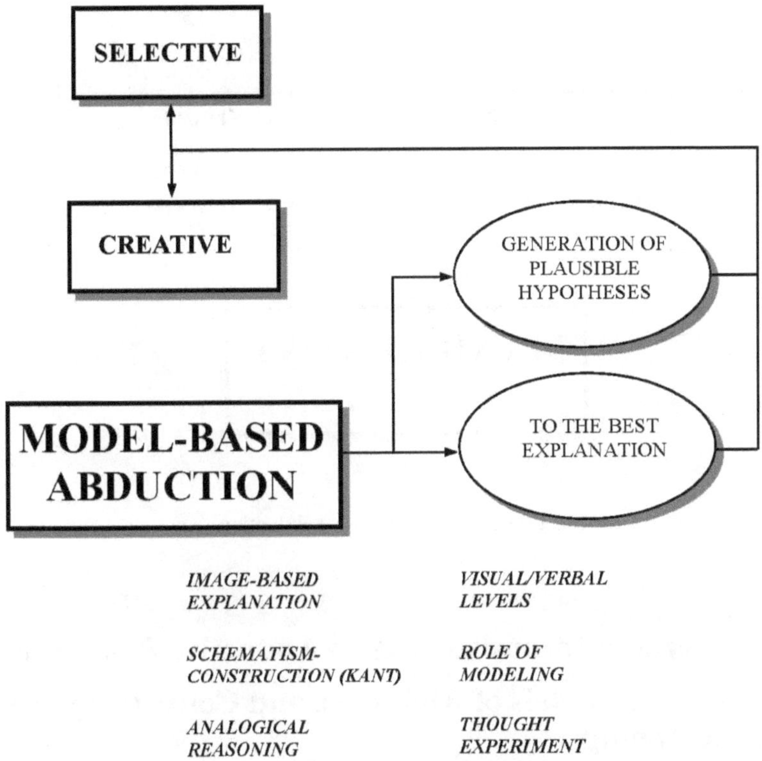

Figure 4. Model-based abduction

duced a pictorial representation that was to play an important role in Maxwell's construction of the quantitative field concept. Use of analogy and imagery in ordinary and scientific problem solving is very complex. Nevertheless, we may observe in many cases all the features of a *productive, creative* mapping, where such "transfer of knowledge" is essential to the development of a new concept. Imagery representations appear to function analogically. The value of an imagery representation is that it makes some structural relations immediately evident.

Visual abduction, but also many kinds of abductions involving analogies, diagrams, thought experimenting, etc., can be called *model-based* (cf. Figure 4). I called *manipulative* that kind of abduction (cf. Figure 5) that involves manipulations of external "mediators" and representations like in the case of classical geometrical reasoning (constructions) and in scientific experiments (construals).[2]

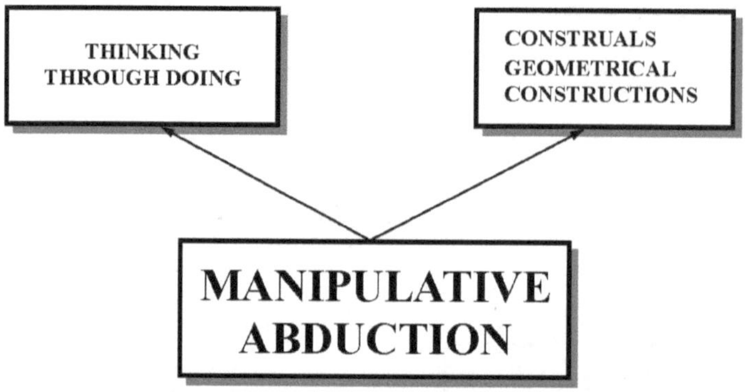

Figure 5. Manipulative abduction

3. Governing Inconsistencies in Abductive Reasoning

3.1 Formal Models of Abductive and Consistency-Based Reasoning

Many attempts have been made to model abduction by developing some formal tools in order to illustrate its computational properties and the relationships with the different forms of deductive reasoning (Bylander et al., 1991; Konolige, 1992; Levesque, 1989; Reiter, 1987; Shanahan, 1989; Reiter and de Kleer, 1987). Some of these formal models of abductive reasoning, for instance Boutilier and Becher, 1995, are based on the theory of the *epistemic state* of an agent (Alchourrón et al., 1985; Gärdenfors, 1988), where the epistemic state of an individual is modeled as a consistent set of beliefs that can change by expansion and contraction[3]. We shall discuss the nature of the kinds of inconsistencies captured by these formalisms and show how they do not adequately account for some roles played by anomalies, conflicts, and contradictions in many forms of explanatory reasoning.

Deductive models of abduction may be characterized as follows (see Figure 6): an explanation for β relative to background theory T will be any α that,

[2]Model-based reasoning and the so-called *constructive modeling* and *generic modeling* in scientific discovery are illustrated by Nersessian in her 1995 and in Nersessian et al., 1997. On manipulative abduction and what I call epistemic mediators in manipulative reasoning, see Magnani, 2001, 2002.

[3]Levi's theory of suppositional reasoning is also related to the problem of "belief change"—Levi, 1996.

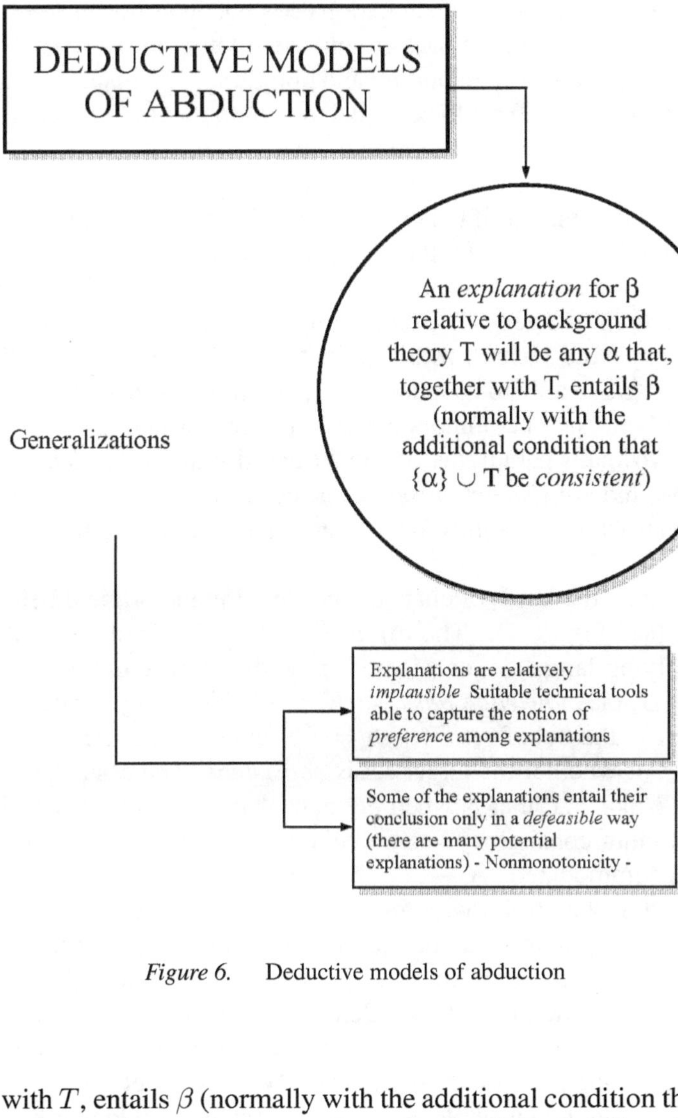

Figure 6. Deductive models of abduction

together with T, entails β (normally with the additional condition that $\{\alpha\} \cup T$ be consistent). Such theories are usually generalized in many directions: first of all by showing that explanations entail their conclusions only in a *defeasible* way (there are many potential explanations), thus joining the whole area of so-called nonmonotonic logic or of probabilistic treatments; second, by trying to show how some of the explanations are relatively implausible, elaborating suitable technical tools (for example in terms of modal logic) able to capture the notion of preference among explanations. Hence, we may require that an explanation makes the observation simply sufficiently probable (Pearl, 1988) or that the explanations that are more likely will be the "preferred" explanations: the involvement of a cat in breaking the glass is less probable than the effect of wind. Finally, the deductive model of abduction does not authorize us to explain facts that are inconsistent with the background theory notwithstanding the fact that these explanations are very important and ubiquitous, for instance in diagnostic applications, where the facts to be explained contradict the expectation that the system involved is working according to specification.

Boutilier and Becher (1995) provide a formal account of the whole question in terms of belief revision: if believing A is sufficient to induce belief in B, then A (epistemically) *explains* B; the situation can be semantically illustrated in terms of an ordering of plausibility or normality which is able to represent the epistemic state of an agent. The conflicting observations will require explanations that compel the agent to withdraw its beliefs (hypotheses), and the derived conditional logic is able to account for explanations of facts that *conflict* with the existing beliefs. The authors are able to reconstruct, within their framework, the two main paradigms of model-based diagnosis, *abductive* (Poole, 1988, 1991), and *consistency-based* (de Kleer et al., 1990; Reiter, 1987), providing an alternative semantics for both in terms of a plausibility ordering over possible worlds.

Let us resume the kinds of change considered in the original belief revision framework (see Figure 7). The *expansion* of a set of beliefs K taken from some underlying language, considered to be the closure of some finite set of premises KB, or *knowledge base*, so $K = Cn(KB)$, by a piece of new information A is the belief set $K + A = Cn(K \cup A)$. The addition happens "regardless" of whether the larger set is *consistent*. The case of *revision* happens when $K \models \neg A$, that is when the new A is *inconsistent* with K and we want to maintain consistency: some beliefs in K must be withdrawn before A can be accommodated: $K \dot{-} A$. The problem is that it is difficult to detect which part of K has to be withdrawn. The least "entrenched" beliefs in K should be withdrawn and A added to the "contracted" set of beliefs. The loss of information has to be as small as possible so that "no belief is given up unnecessarily" (Gärdenfors, 1988). Hence, *inconsistency resolution* in a belief revision framework is captured by the concept of revision. Another way of belief change is the process of *contraction*. When a belief set K is contracted by A, the resulting belief set $K \dot{+} A$ is such that A is no longer held, without adding any new fact.

After having explained the distinction between predictive explanations and "might" explanations, that merely allow an observation, and do not predict it, Boutilier and Becher show in the cited article how model-based diagnoses can be accounted for in terms of their new formal model of belief revision.

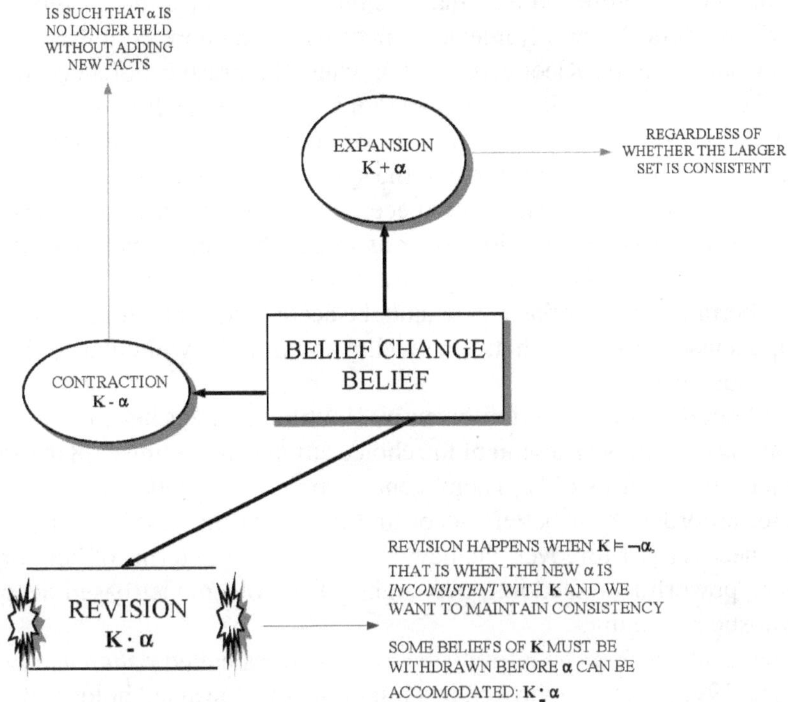

Figure 7. Belief revision framework

The *abductive* model-based reasoning[4] (Poole, 1988, 1991; Brewka, 1989) illustrated by some models, such as Poole's Theorist, allows many possible explanations, weak and predictive (so presenting a paraconsistent behavior: a non-predictive hypothesis can explain both a proposition and its negation). This old model, embedded in the new formal framework, acquires the possibility of discriminating certain explanations as preferred to others. Reiter's *consistency-based* diagnosis (Reiter, 1987) is devoted to ascertain why a correctly designed system is not working according to its features. Because certain components may fail, the system description also contains some abnormality predicates (the absence of them will render the description inconsistent with an observation of an incorrect behavior). The consistency-based diagnosis concerns any set of components whose abnormality makes the observation consistent with the

[4]Please distinguish here the technical use of the attribute *model-based* from the epistemological one I introduced in the previous section.

description of the system. A principle of parsimony is also introduced to capture the idea of preferred explanations/diagnoses. Since the presence of fault models renders Reiter's framework incorrect, new more complicated notions are introduced in de Kleer et al., 1990, where the presence of a complete fault model ensures that predictive explanations may be given for "every" abnormal observation. Without any description of correct behavior any observation is consistent with the assumption that the system works correctly. Hence, a complete model of correct behavior is necessary if we want the consistency-based diagnosis to be useful. The idea of consistency that underlies this kind of diagnostic reasoning is the following: any inconsistency (anomalous observation) is an aberrant behavior that can usually be accounted for by finding some set of components of a system that, if behaving abnormally, will entail or justify the actual observation.

Without doubt the solution given by Boutilier and Becher furnishes a more satisfying qualitative account of the choice among competing explanations than Gärdenfors' in terms of "epistemic entrenchment"[5], which tries to capture the idea of an ordering of beliefs according to our willingness to withdraw them when necessary. Moreover, the new formal account in terms of belief revision is very powerful in shedding new light on the old model-based accounts of diagnostic reasoning.

The framework of belief revision is sometimes called *coherence approach* (Doyle, 1992). In this approach it is important that the agent holds some beliefs just as long as they are consistent with the agent's remaining beliefs. Inconsistent beliefs do not describe any world, and so are unproductive; moreover, the changes must be epistemologically conservative in the sense that the agent maintains as many of its beliefs as possible when it adjusts its beliefs to the new information. It is contrasted to the *foundations approach*, according to which beliefs change as the agent adopts or abandons satisfactory reasons (or justifications). This approach is exemplified by the well-known "reason maintenance systems" (RMS) or "truth maintenance systems" (TMS) (Doyle, 1979), elaborated in the area of artificial intelligence to cooperate with an external problem solver. In this approach the role of inconsistencies is concentrated on the negations able to invalidate justifications of beliefs; moreover, as there are many similarities between reasoning with incomplete information and acting with inconsistent information, the operations of RMS concerning revision directly involve logical consistency, seeking to solve a conflict among beliefs. The operations of *dependency-directed backtracking* (DDB) are devoted to this aim: RMS informs DDB whenever a contradiction node (for instance a set of beliefs) becomes believed, then DDB attempts to remove reasons and premises, only to

[5]Which of course may change over time or with the state of belief.

defeat nonmonotonic assumptions: "If the argument for the contradiction node does not depend on any of these (i.e., it consists entirely of monotonic reasons), DDB leaves the contradiction node in place as a continuing belief" (Doyle, 1992, p. 36), so leaving the conflicting beliefs intact if they do not depend on defeasible assumptions, and presenting a paraconsistent behavior.

Both in the coherence and foundations approach the changes of state have to be *epistemologically conservative*: as already said above the agent maintains as many of its beliefs as possible when it adjusts its beliefs to the new information, thus following Quine's idea of "minimum mutilation" (Quine, 1979). We have now to notice some limitations of the formal models in accounting for other kinds of inconsistencies embedded in many reasoning tasks. Important developments in the field of logical models of abduction—also touching some related problems in artificial intelligence (AI) and devoted to overcome the limitations above—are illustrated in Flach and Kakas, 2000 and in Gabbay and Kruse, 2000; Gabbay and Woods, 2005; Gabbay and Woods, 2006; Meheus et al., 2002; Meheus and Batens, 2006. See also the papers contained in Magnani and Nersessian, 2002 and in Magnani, 2006b.

3.2 More Conflictual Features, Creative Settings, Coherence

If we want to deal with the nomological and most interesting creative aspects of abduction we are first of all compelled to consider the whole field of the growth of scientific knowledge cited above. We have anticipated that abduction has to be an inference permitting the derivations of *new* hypotheses and beliefs. Some explanations consist of certain facts (initial conditions) and universal generalizations (that is scientific laws) that deductively entail a given fact (observation), as showed by Hempel in his *covering law model* of scientific explanation (Hempel, 1966). If T is a theory illustrating the background knowledge (a scientific or common sense *theory*) the sentence α explains the fact (observation) β just when $\{\alpha\} \cup T \models \beta$. It is difficult to govern the question involving nomological and causal aspects of abduction and explanation in the framework of the belief revision illustrated in the previous section: we would have to deal with a kind of belief revision that permits us to alter a theory with new conditionals.

We may also see belief change from the point of view of *conceptual change*, considering concepts either cognitively, like mental structures analogous to data structures in computers, or, epistemologically, like abstractions or representations that presuppose questions of justification. Belief revision—even if extended by formal accounts such as illustrated above[6]—is able to represent

[6]Or developed by others, see, for example, Katsuno and Mendelzon, 1992; Cross and Thomason, 1992.

cases of conceptual change such as adding a new instance, adding a new weak rule, adding a new strong rule (see Thagard, 1992, pp. 34–39, for details), that is, cases of addition and deletion of beliefs, but fails to take into account cases such as adding a new part-relation, adding a new kind-relation, adding a new concept, collapsing part of a kind-hierarchy, reorganizing hierarchies by branch jumping and tree switching, in which there are reorganizations of concepts or redefinitions of the nature of a hierarchy. These last cases are the most evident changes occurring in many kinds of creative reasoning, for example in science. Related to some of these types of conceptual change are different varieties of inconsistencies (see Figure 8).

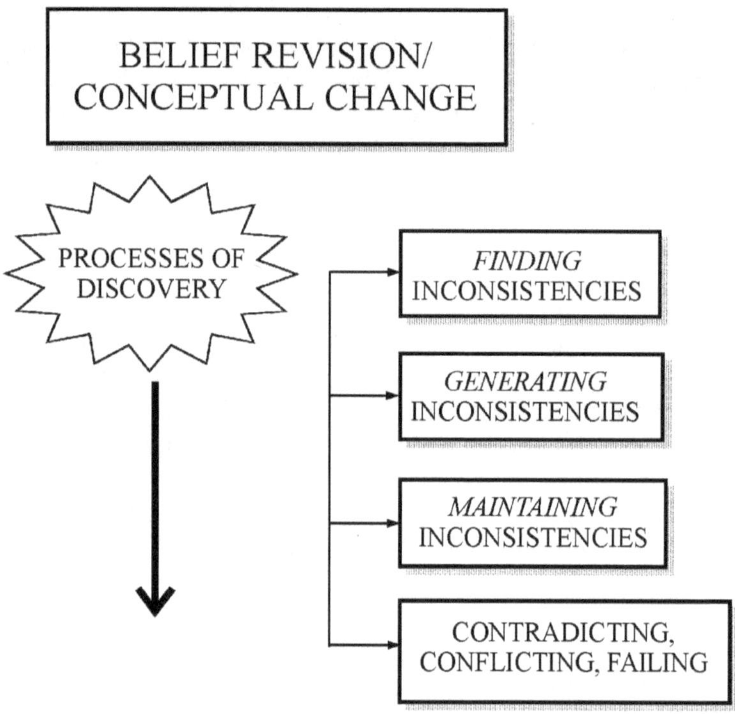

Figure 8. Conceptual chance and inconsistencies

Finding Inconsistencies: Empirical and Conceptual Anomalies. It may be said that the logical accounts of abduction described above certainly illustrate much of what is important in abductive reasoning, especially the objective of selecting a set of hypotheses (diagnoses, causes) that are able to dispense good (preferred) explanations of data (observations), but fail in accounting for many cases of explanations occurring in science or in everyday reasoning. For example they do not capture

1 the role of statistical explanations, where what is explained follows only probabilistically and not deductively from the laws and other tools that do the explaining;

2 the sufficient conditions for explanation;

3 the fact that sometimes the explanations consist of the application of schemas that fit a phenomenon into a pattern without realizing a deductive inference;

4 the idea of the existence of high-level kinds of *creative* abductions I cited above;

5 the existence of model-based abductions (for instance visual and diagrammatic);

6 the fact that explanations usually are not complete but only furnish *partial* accounts of the pertinent evidence (see Thagard and Shelley, 1997).

Moreover, the logical accounts of abduction certainly elucidate many kinds of inconsistency government, which nevertheless reduce to the act of finding contradictions able to generate the withdrawal of some hypotheses, beliefs, reasons, etc.: these contradictions always emerge at the level of data (observations), and consistency is restored at the theoretical level.[7] This view may distract from important aspects of other kinds of reasoning that involve intelligent abductive performances.

For example, *empirical anomalies* are not alone in generating impasses, there are also so-called conceptual anomalies. In science, first and foremost, empirical anomaly resolution involves the localization of the problem at hand within one or more constituents of the theory. It is then necessary to produce one or more new hypotheses to account for the anomaly, and finally, these hypotheses need to be evaluated so as to establish which one best satisfies the criteria for theory justification. Hence, anomalies require a change in the theory, yet once the change is successfully made, anomalies are no longer anomalous but in fact are now resolved. General strategies for anomaly resolution, as well as for producing new ideas and for assessing theories, have been studied by Darden (1991).

The so-called *conceptual problems* represent a particular form of anomaly. In addition, resolving conceptual problems may involve satisfactorily answering

[7] We have to remember that the logical models in some cases exhibit a sort of paraconsistent behavior.

questions about the nature of theoretical entities. Nevertheless such conceptual problems do not arise directly from data, but from the nature of the claims in the principles or in the hypotheses of the theory. It is far from simple to identify a conceptual problem that requires a resolution, since, for example, a conceptual problem concerns the adequacy or the ambiguity of a theory, and yet also its incompleteness or (lack of) evidence. In Magnani, 1997 I present some examples derived from the historical discovery of non-Euclidean geometries which illustrate the relationships between strategies for anomaly resolution and explanatory and productive visual thinking: I consider how visual thinking is relevant to hypothesis formation and scientific discovery and explore the first epistemological and cognitive features of what I described above as *visual abduction.*

The fact that inconsistencies may occur also at the theoretical level is further emphasized if we consider that in science or in legal reasoning (Thagard, 1992), hypotheses are mainly *layered.* Hence, the organization of hypotheses is more complex than the one illustrated in previous formal models, and abduction is not only a matter of mapping from sets of hypotheses to a set of data. In many abductive settings there are hypotheses that explain other hypotheses so that the selection or creation of explanations is related to these relationships. This kind of hierarchical explanations has also been studied in the area of probabilistic belief revision (Pearl, 1988).

Generating Inconsistencies by Radical Innovation. The case of conceptual change such as adding a new part-relation, adding a new kind-relation, adding a new concept, collapsing part of a kind-hierarchy, reorganizing hierarchies by branch jumping and tree switching, in which there are reorganizations of concepts or redefinitions of the nature of a hierarchy are the most evident changes occurring in many kinds of *creative reasoning*, for instance in the growth of scientific knowledge.

When a scientist introduces a new hypothesis, especially in the field of the natural sciences, he is interested in the potential rejection of an old theory or of an old knowledge domain. Consistency requirements we described in the framework of deductive models, governing hypothesis withdrawal in various ways, would arrest further developments of the new abduced hypothesis. In the scientist's case there is not the deletion of the old concepts, but rather the *coexistence* of two rival and competing views. Consequently we have to consider this competition as a form of epistemological, and not logical inconsistency. For instance two scientific theories are conflicting because they compete in explaining shared evidence.

The problem has been studied in Bayesian terms but also in connectionist ones, using the so-called theory of explanatory coherence (Thagard, 1992), which deals with the epistemological (but sometimes pragmatical) reasons for accepting a whole set of explanatory hypotheses conflicting with another one. In some cognitive settings, such as the task of comparing a set of hypotheses and beliefs incorporated in a scientific theory with the one of a competing theory, we have to consider a very complex set of criteria (to ascertain which composes the best explanation), that goes beyond mere simplicity or explanatory power. The minimality criteria included in some of the formal accounts of abduction, or the idea of the choice among preferred models cited above, are not sufficient to illustrate more complicated cognitive situations.

Maintaining Inconsistencies. As noted in the previous subsection, when we create or produce a new concept or belief that competes with another one, we are compelled to maintain the derived inconsistency until the possibility of rejecting one of the two becomes feasible. Other cognitive and epistemological situations present a sort of paraconsistent behavior: a typical kind of *consistency maintenance* is the well-known case of scientific theories that face anomalies. As noted above, explanations are usually not complete but only furnish partial accounts of the pertinent evidence: not everything has to be explained.

Newtonian mechanics is forced to cohabit with the anomaly of the perihelion of Mercury until the development of the theory of relativity, but it also has to stay with its false prediction about the motion of Uranus. In diagnostic reasoning too, it is necessary to make a diagnosis even if many symptoms are not explained or remain mysterious. In this situation we again find the similarity between reasoning in the presence of inconsistencies and reasoning with incomplete information already stressed. Sometimes scientists may generate so-called *auxiliary hypotheses* (Lakatos, 1970), justified by the necessity of overcoming these kinds of inconsistencies: it is well-known that the auxiliary hypotheses are more acceptable if able to predict or explain something new (the making of the hypothesis of the existence of another planet, Neptune, was a successful way—not an *ad hoc* manoeuvre—of eliminating the anomaly of the cited false prediction).

Contradicting, Conflicting, Failing. Considering the *coherence* of a conceptual system as a matter of the simultaneous satisfaction of a set of positive and negative constraints leads to the *connectionist models* (also in computational terms) of coherence. In this light logical inconsistency becomes a relation that furnishes a *negative* constraint and entailment becomes a relation that provides a *positive* constraint. For example, as already noted, some hypotheses are inconsistent when they simply compete, when there are some pragmatic incompatibility relations, when there are incompatible ways of combining images, etc. (Thagard and Verbeurgt, 1998).

From the viewpoint of the connectionist model of coherence, situations are allowed in which there is a set of accepted concepts containing an inconsistency (see previous subsection), for example in the case of anomalies: the system at hand may at any rate have a maximized coherence, when compared to another system. Moreover, "another interesting case is the relation between quantum theory and general relativity, two theories which individually possess enormous explanatory coherence. According to the eminent mathematical physicist Edward Witten, 'the basic problem in modern physics is that these two pillars are incompatible [. . .]'. Quantum theory and general relativity may be incompatible, but it would be folly given their independent evidential support to suppose that one must be rejected" (ibid.).

Hypotheses may be *unfalsifiable.* In this case it is impossible to find a contradiction in some area of the conceptual systems in which they are incorporated. Notwithstanding this fact, it is sometimes necessary to construct ways of rejecting the unfalsifiable hypothesis at hand by resorting to some *external* forms of negation, external because we want to avoid any arbitrary and subjective elimination, which would be rationally or epistemologically unjustified.

In the following section we will consider a kind of "weak" hypothesis that is hard to negate and the ways for making it easy. In these cases the subject can *rationally* decide to withdraw his hypotheses even in contexts where it is *impossible* to find *"explicit" contradictions*; moreover, thanks to the new information reached simply by finding this kind of negation, the subject is *free* to abduce new hypotheses. I will explore whether *negation as failure* can be employed to model hypothesis withdrawal in Poincaré's conventionalism of the principles of physics, showing how conventions can be motivationally abandoned.

4. Withdrawing Unfalsifiable Hypotheses

4.1 Negation as Failure in Query Evaluation

There is a kind of negation, studied by researchers into logic programming, which I consider to be very important also from the epistemological point of view: *negation as failure.* It is active as a "rational" process of withdrawing previously-imagined hypotheses in everyday life, but also in certain subtle kinds of diagnostic and epistemological settings. Contrasted with classical negation, with the double negation of intuitionistic logic, and with the philosophical concept of *Aufhebung*, negation as failure shows how a subject can decide to withdraw his hypotheses, while maintaining the *rationality* of his reasoning, in contexts where it is impossible to find explicit contradictions; as stated above, thanks to the new information reached simply by finding this kind of negation, the subject is *free* to form new hypotheses.

The statements of a logical data base are a set of Horn clauses which take the form:

$$R(t_1, \ldots, t_n) \leftarrow L_1 \wedge L_2 \wedge \ldots \wedge L_m$$

($m \geq 0, n \geq 0$, where $R(t_1, \ldots, t_n)$—conclusion—is the distinguished positive literal[8] and $L_1 \wedge L_2 \wedge \ldots \wedge L_m$—conditions—are all literals, and each free variable is implicitly universally quantified over the entire implication). In more conventional notation this would be written as the disjunction

$$R(t_1, \ldots, t_n) \vee \neg L_1 \vee \neg L_2 \vee \ldots \vee \neg L_m$$

where any other positive literal of the disjunctive form would appear as a negated precondition of the previous implication.

Let us consider a special query evaluation process for a logical data base that involves the so-called *negation as failure* inference rule (Clark, 1978). We can build a Horn clause theorem prover augmented with this special inference rule, such that we are able to infer $\neg P$ when every possible proof of P *fails*.

We know that a relational data base only contains information about *true* instances of relations. Even so, many queries involve negation and we can answer them by showing that certain instances are *false*. For example, let's consider this simple case: to answer a request for the name of a student not taking a particular course, C, we need to find a student, S, such that the instance (atomic formula) $Takes(S, C)$ is false. For a logical data base, where an atomic formula which is not explicitly given may still be implied by a general rule, the assumption is that an atomic formula is false if we *fail* to prove that it is true. To prove that an atomic formula P is *false* we do an exhaustive search for a proof of P. If *every* possible proof of P fails, we can infer $\neg P$. The well-known PROLOG programming language (Roussel, 1975) uses this method of manipulating negation.

We have to deal with a proof such as the following:

from proving $\nvdash P$ infer $\vdash \neg P$

where the "proof that P is not provable" (Clark, 1978, p. 120) is the *exhaustive* but *unsuccessful search* for a proof of P. Here the logical symbol $\neg$ acquires the new meaning of "fail to prove" (see Figure 9).

Clark proposes a query evaluation algorithm based essentially on ordered linear resolution for Horn clauses (SLD) augmented by the negation as failure inference rule "$\neg P$ may be inferred if every possible proof of P fails" (SLDNF).[9]

What is the semantic significance of this kind of negation? Can we interpret a failed proof of P as a *valid* first order inference that P is false? Clark's response

---

[8] A *literal* is an atomic formula or the negation of an atomic formula.

[9] The links between negation as failure, completed data bases (Clark, 1978), and the closed world assumption (Shepherdson, 1984, 1988) have been studied in great detail. A survey can be found in Lloyd, 1987.

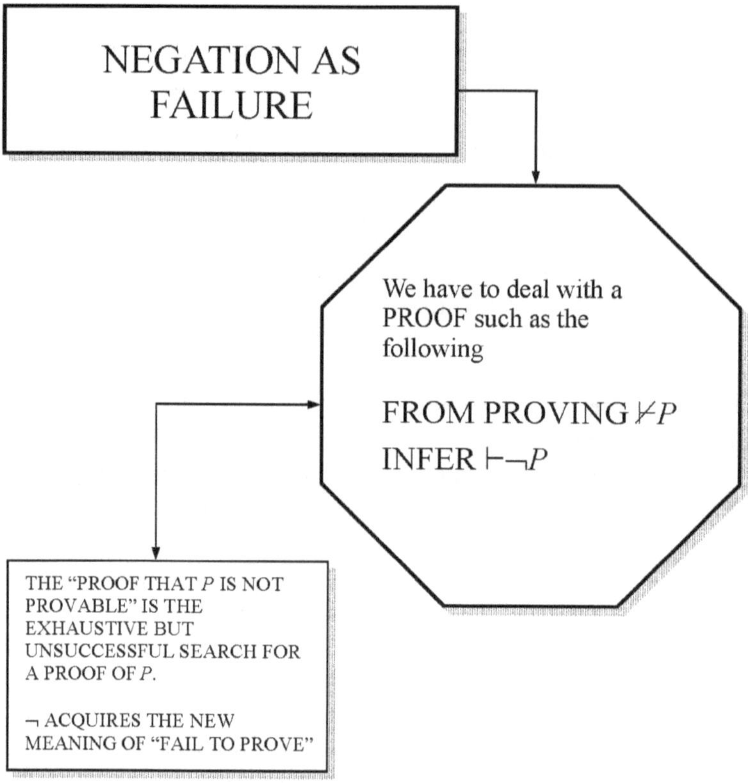

Figure 9. Negation as failure

resorts to reconciling negation as failure with its truth functional semantics. If we can demonstrate that every failed attempt to prove P using the data base of clauses B, is in effect a proof of $\neg P$ using the *completed*[10] data base $C(B)$, then "negation as failure" is a derived inference rule for deductions from $C(B)$. The explicit axioms of equality and completion laws are therefore necessary at the object level in order to simulate failure of the matching algorithm at the meta-level. A negated literal $\neg P$ will be evaluated by recursively entering the algorithmic query evaluator (as an ordered linear resolution proof procedure, as stated above) with the query P. If every possible path for P ends in failure

---

[10]The notion of *data base completion* can be found in Clark, 1978 and in all textbooks on logic for computer science.

(failure proofs that can be nested to any depth), we return with $\neg P$ evaluated as true.

Clark (1978) has shown that for every meta-language proof of $\neg P$ obtained by a Horn clause theorem prover (query evaluation) augmented with negation as failure there exists a structurally similar object-language proof of $\neg P$. He has proved that a query evaluation with the addition of negation as failure will only produce results that are implied by first order inference from the completed data-base, that is, the evaluation of a query should be viewed as a "deduction" from the completed data base *correctness of query evaluation.* Consequently negation as failure is a sound rule for deductions from a completed data base.

Although the query evaluation with negation as failure process is in general *not complete*, its main advantage is the efficiency of its implementation. There are many examples in which the attempt to prove neither succeeds nor fails, because it goes into a loop. To overcome these limitations it is sufficient to impose constraints on the logical data base and its queries, and add loop detectors to the Horn clause problem solver: by this method the query evaluation process is guaranteed to find each and every solution to a query.

However, because of the undecidability of logic (Church, 1936), no query evaluator can identify all cases in which a goal in unsolvable. A best theorem prover does not exist and there are no limitations on the extent to which a problem solver can improve its ability to detect loops and to establish negation as failure.

In the next subsection I will consider some aspects dealing with Poincaré's famous conventionalism of the principles of physics and the possibility of negating conventions.

4.2 Withdrawing "Conventions"

An extension of Poincaré's so-called *geometric conventionalism*, according to which the choice of a geometry is only justifiable by considerations of simplicity, in a psychological and pragmatic sense ("commodisme"), is the *generalized conventionalism*, expressing the conventional character of the principles of physics: "The principles of mathematical physics (for example, the principle of conservation of energy, Hamilton's principle in geometrical optics and in dynamics, etc.) systematize experimental results usually achieved on the basis of two (or more) rival theories, such as the emission and the undulation theory of light, or Fresnel's and Neumann's wave theories, or Fresnel's optics and Maxwell's electromagnetic theory, etc. They express the common empirical content as well as (at least part of) the mathematical structure of such rival theories and, therefore, can (but need not) be given alternative theoretical interpretations" (Giedymin, 1982, pp. 27–28).

From the epistemological point of view it is important to stress that the conventional principles usually survive the demise of theories and are therefore responsible for the continuity of scientific progress. Moreover, they are not empirically falsifiable; as stated by Poincaré in *Science and Hypothesis*:

> The principles of mechanics are therefore presented to us under two different aspects. On the one hand, they are truths founded on experiment, and verified approximately as far as almost isolated systems are concerned; on the other hand they are postulates applicable to the whole of the universe and regarded as rigorously true. If these postulates possess a generality and a certainty which the experimental truths from which they were derived lack, it is because they reduce in final analysis to a simple convention that we have a right to make, because we are certain beforehand that no experiment can contradict it. This convention, however, is not absolutely arbitrary; it is not the child of our caprice. We admit it because certain experiments have shown us that it will be convenient, and thus is explained how experiment has built up the principles of mechanics, and why, moreover, it cannot reverse them. (Poincaré, 1902, pp. 135–136)

The conventional principles of mechanics derive from experience, as regards their "genesis", but cannot be falsified by experience because they contribute to "constitute" the experience itself, in a proper Kantian sense. The experience has only suggested their adoption because they are *convenient*: there is a precise analogy with the well-known case of geometrical conventions, but also many differences, which pertain to the "objects" studied.[11]

Poincaré seeks also to stress that geometry is more abstract than physics, as is revealed by the following speculations about the difficulty of "tracing artificial frontiers between the sciences":

> Let it not be said that I am thus tracing artificial frontiers between the sciences; that I am separating by a barrier geometry properly so called from the study of solid bodies. I might just as well raise a barrier between experimental mechanics and the conventional mechanics of general principles. Who does not see, in fact, that separating these two sciences we mutilate both, and that what will remain of the conventional mechanics when it is isolated will be but very little, and can in no way be compared with that grand body of doctrine which is called geometry. (Poincaré, 1902, pp. 137–138)

I believe that the meaning of this passage refers primarily to the fact that physics cannot be considered completely conventional because we know that the conventional "principles" are derived from the "experimental laws" of "experimental mechanics", and then absolutized by the "mind". Second, Poincaré wants to demonstrate how geometry is more abstract than physics: geometry does not

[11]The conventional principles of mechanics should not be confused with geometrical conventions: "The experiments which have led us to adopt as more convenient the fundamental conventions of mechanics refer to bodies which have nothing in common with those that are studied by geometry. They refer to the properties of solid bodies and to the propagation of light in a straight line. These are mechanical, optical experiments" (Poincaré, 1902, p. 137), they are not, Poincaré immediately declares, "*des expériences de géométrie*" (ibid.).

require a rich experimental reference as physics does, geometry only 'requires' that experience regarding its genesis and as far as demonstrating that it is the most convenient is concerned. Here we are very close to Kant's famous passage about the *synthetical a priori* character of the judgments of (Euclidean) geometry, and of the whole of mathematics: "Mathematics presents the most splendid example of the successful extension of pure reason, without the help of experience" (Kant, 1929, A712-B740, p. 576).

Even when separated from the reference to solid bodies, Euclidean geometry maintains all its conceptual pregnancy, as a convention that, in a proper Kantian sense, "constitutes" the ideal solid bodies themselves. This is not the case of the conventional principles of mechanics when separated from experimental mechanics: "what will remain of the conventional mechanics [...] will be very little" if compared "with that grand body of doctrine which is called geometry".

Poincaré continues:

> Principles are conventions and definitions in disguise. They are, however, derived from experimental laws, and these laws have, so to speak, been erected into principles to which our mind attributes an absolute value. Some philosophers have generalized far too much. They have thought that the principles were the whole of science, and therefore that the whole of science was conventional. This paradoxical doctrine, which is called nominalism, cannot stand examination. How can a law become a principle? (Poincaré, 1902, p. 138)

If the experimental laws of experimental physics are the source of the conventional principles themselves, conventionalism escapes nominalism.

As stated at the beginning of this subsection, conventional principles survive the demise (falsification) of theories in such a way that they underlie the incessant spectacle of scientific revolutions. Underlying revolutions of physics, conventional principles guarantee the historicity and the growth of science itself. Moreover the conventional principles surely imply "*firstly*, that there has been a *growing tendency* in modern physics to *formulate and solve physical problems* within *powerful, and more abstract, mathematical systems of assumptions* [...]; *secondly*, the role of conventional principles has been growing and *our ability to discriminate experimentally between alternative abstract systems* which, with a great approximation, save the phenomena *has been diminishing* (by comparison to the testing of simple conjunctions of empirical generalizations)" (Giedymin, 1982, p. 28).

Up to now we have considered in detail how the conventional principles guarantee the revolutionary changes of physics and why they cannot be considered arbitrary, being motivated by the "experimental laws" of the "experimental physics", that is by experience. Although arbitrary and conventional the conventional principles too can be substituted by others. This is the main problem treated by Poincaré in the last passages of Chapter IX, "The Future of Mathematical Physics", in *The Value of Science*. Already the simple case of "linguistic" changes in science "suffices to reveal generalizations not before suspected"

(Poincaré, 1905, p. 78). By means of the new discoveries scientists arrive at a point where they are able to "admire the delicate harmony of numbers and forms; they marvel when a new discovery opens to them an unexpected perspective" (Poincaré, 1905, pp. 75–76), a new perspective that is always provisional, fallible, open to further confirmations or falsifications when compared to rival perspectives. We have seen how the conventional principles of physics guarantee this continuous extension of experience thanks to the various perspectives and forms expressed by experimental physics. However, because conventional, "no experiment can contradict them". The experience only suggested the principles, and they, since absolute, have become constitutive just of the empirical horizon common to rival experimental theories.

Poincaré observes:

> Have you not written, you might say if you wished to seek a quarrel with me—have you not written that the principles, though of experimental origin, are now unassailable by experiment because they have become conventions? And now you have just told us that the most recent conquests of experiment put these principles in danger. Well, formerly I was right and to-day I am not wrong. Formerly I was right, and what is now happening is a new proof of it. (Poincaré, 1905, p. 109)

Poincaré appeals to a form of weak negation, just as Freud (see below, footnote 12) did when dealing with the problem of withdrawing constructions. Let us follow the text. To pursue his point, Poincaré illustrates the attempts to reconcile the "calorimetric experiment of Curie" with the "principle of conservation of energy":

> This has been attempted in many ways; but there is among them one I should like you to notice; this is not the explanation which tends to-day to prevail, but it is one of those which have been proposed. It has been conjectured that radium was only an intermediary, that it only stored radiations of unknown nature which flashed through space in every direction, traversing all bodies, save radium, without being altered by this passage and without exercising any action upon them. Radium alone took from them a little of their energy and afterward gave it out to us in various forms. (Poincaré, 1905, pp. 109–110)

At this point Poincaré resolutely asserts: "What an advantageous explanation, and how convenient! First, it is unverifiable and thus irrefutable. Then again it will serve to account for any derogation whatever to Mayer's principle; it answers in advance not only the objection of Curie, but all the objections that future experimenters might accumulate. This new and unknown energy would serve for everything" (p. 110). Now Poincaré can show how this *ad hoc* hypothesis can be identified with the non falsifiability of the conventional principle of the conservation of energy:

> This is just what I said, and therewith we are shown that our principle is unassailable by experiment. But then, what have we gained by this stroke? The principle is intact, but thenceforth of what use is it? It enabled us to foresee that in such

> and such circumstance we could count on such total quantity of energy; it limited us; but now that this indefinite provision of new energy is placed at our disposal, we are no longer limited by anything. (Poincaré, 1905, p. 110)

Finally, Poincaré's argumentation ends by affirming negation as failure: "and, as I have written in 'Science and Hypothesis', if a principle ceases to be fecund, experiment without contradicting it directly will nevertheless have condemned it" (ibid.).

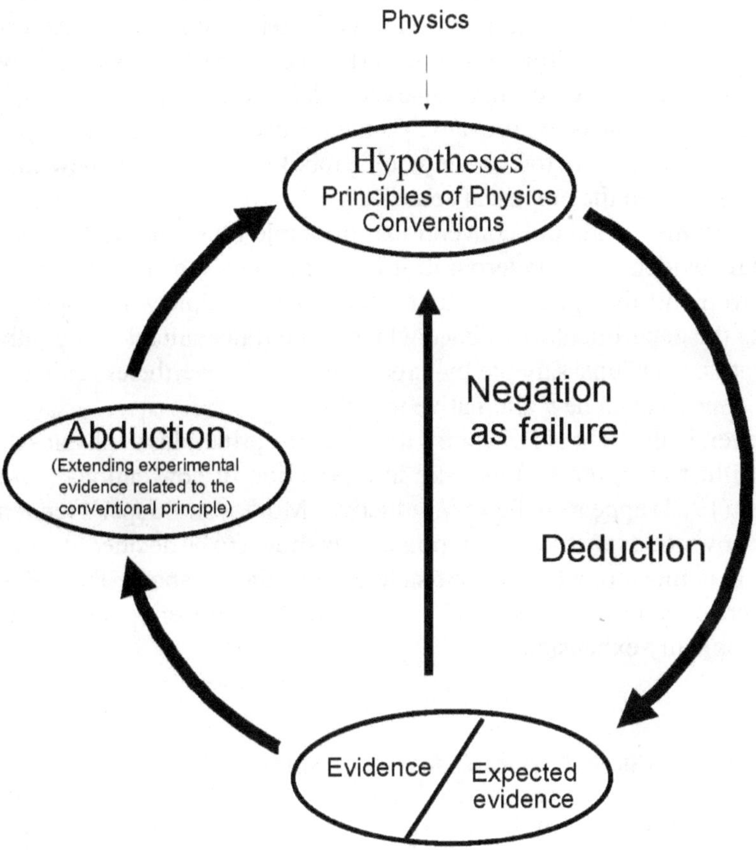

Figure 10. Withdrawing conventions

Let us now analyze this situation from the epistemological point of view (see Figure 10): the conventional principle has to be withdrawn when it "ceases to be fecund", or when it seems that we have failed to prove it. In the computational case, negation as failure is achieved by suitable algorithms related to the knowledge that is handled (see above, Subsection 4.1). Remember that for a logic data base the assumption is that an atomic formula is false if we *fail* to prove that it is true. More clearly: as stated above, every conventional principle, suitably underlying some experimental laws, generates expectations with regard to the subsequent evidences of nature. We consider as proof of a conventional principle the fact that we can increasingly extend and complete the experimental laws related to it, adding the new (expected) evidence that "emerges" from the experimental research. If, after a finite period of time, nature does not provide this new "evidence" that is able to increase the fecundity of the conventional principle, this *failure* leads to its withdrawal: "experiment without contradicting it directly will nevertheless have condemned it". Analogously to the Freudian case I have analyzed elsewhere (Magnani, 2005)[12], the "proof that a principle is not provable" is the unsuccessful search for a proof of the principle itself. Here too, the logical symbol $\neg$ acquires the new meaning of "fail to prove" in the empirical sense.

Let us resume: if the old conventional principle does not produce new experimental "evidence" to underpin it, it is legitimate to abandon the principle, when convenient: the opportunity to reject the old principle will happen just by exploiting the experimental evidence which, even if not suitable for contradicting it (that is, it is "unassailable by experiment"), is nevertheless suitable as a basis for conceiving a new alternative principle.

Moreover, in the light of Poincaré's theory of the principles of physics that we have just illustrated, the nominalistic interpretation of conventionalism given by Popper (1963) appears to be very reductive. Moreover, Popper's tendency to identify conventions with *ad hoc* hypotheses is shown to be decidedly unilateral, since, as is demonstrated by the passages, immediately above, the *adhocness* is achieved only in a very special case, when the conventional principle is epistemologically exhausted.

References

Alchourrón, C., Gärdenfors, . P., and Makinson, . P. (1985). On the theory of logic change: partial meet functions for contractions and revision. *Journal of Symbolic Logic*, 50:510–530.

Anderson, A. and Belnap, N. (1975). *Entailment*. Princeton University Press, Princeton.

Boutilier, C. and Becher, V. (1995). Abduction as belief revision. *Artificial Intelligence*, 77:43–94.

[12]Negation as failure can be employed to model hypothesis withdrawal in Freudian analytic reasoning—withdrawing constructions—to explain how the questioned problem of the probative value of clinical findings can be solved. In this representative human and not computational case, negation as failure is played out in the midst of the analyst-analysand interaction, where transference and countertransference are the human epistemological operators and "reagents". Negation as failure is therefore a limitation on the dogmatic and autosuggestive exaggerations of (pathological) countertransference.

Brewka, G. (1989). Preferred subtheories: an extended logical framework for default reasoning. In *Proceedings IJCAI-89*, pages 1043–1048, Detroit, MI.

Bylander, T., Allemang, D., Tanner, M. C., and Josephson, J. R. (1991). The computational complexity of abduction. *Artificial Intelligence*, 49:25–60.

Cantoni, V., editor (1994). *Human and Machine Vision: Analogies and Divergences*. Plenum, New York.

Church, A. (1936). A note on the Entscheidungsproblem. *Journal of Symbolic Logic*, 1:40–41. Correction, ibid., 101-102.

Clark, K. L. (1978). Negation as failure. In Gallaire, H. and Minker, J., editors, *Logic and Data Bases*, pages 119–140. Plenum, New York. (Reprinted in Ginsberg, 1987, pages 311-325.)

Cross, C. and Thomason, R., H. (1992). Conditionals and knowledge-base update. In Gärdenfors, 1992, pages 247–275.

Dalla Chiara, M. L., Doets, K., Mundici, D., and van Benthem, J., editors (1997). *Logic and Scientific Methods*. Kluwer, Dordrecht.

Darden, L. (1991) *Theory Change in Science: Strategies from Mendelian Genetics*. Oxford University Press, New York.

de Kleer, J., Mackworth, A. K., and Reiter, R. (1990). Characterizing diagnoses. In *Proceedings AAAI-90*, pages 324–330, Boston, MA.

Doyle, J. (1979). A truth maintenance system. *Artificial Intelligence*, 12:231–272.

Doyle, J. (1992). Reason maintenance and belief revision: Foundations versus coherence theories. In Gärdenfors, 1992, pages 29–51.

Evans, D. and Patel, V., editors (1992). *Advanced Models of Cognition for Medical Training and Practice*. Springer, Berlin.

Flach, P. and Kakas, A., editors (2000). *Abductive and Inductive Reasoning: Essays on Their Relation and Integration*. Kluwer, Dordrecht.

Gabbay, D. and Kruse, R. (2000). *Abductive Reasoning and Learning*. In Ketner, K. L., editor, *Handbook of Defeasible Reasoning and Uncertainty Management Systems*, vol. 4. Kluwer, Dordrecht.

Gabbay, D. and Woods, J. (2005). *The Reach of Abduction*. North-Holland, Amsterdam. Volume 2 of *A Practical Logic of Cognitive Systems*.

Gabbay, D. and Woods, J. (2006). A formal model of abduction. In Magnani, 2006a.

Gärdenfors, P. (1988). *Knowledge in Flux*. The MIT Press, Cambridge.

Gärdenfors, P., editor (1992). *Belief Revision*. Cambridge University Press, Cambridge.

Giedymin, J. (1982). *Science and Convention. Essays on Henri Poincaré's Philosophy of Science and the Conventionalist Tradition*. Pergamon, Oxford.

Ginsberg, M. L., editor (1987). *Readings in Nonmonotonic Reasoning*. Morgan Kaufmann, Los Altos, CA.

Hempel, C. G. (1966). *Philosophy of Natural Science*. Prentice-Hall, Englewood Cliffs, NJ.

Holyoak, H. J. and Thagard, P. (1996). *Mental Leaps. Analogy in Creative Thought*. The MIT Press, Cambridge, MA.

Josephson, J., Chandrasekaran, B., Smith, J., and Tanner, M. (1986). Abduction by classification and assembly. *PSA 1986*, 1, Philosophy of Science Association:458–470.

Josephson, J. and Josephson, S. G. (1994). *Abductive Inference. Computation, Philosophy, Technology*. Cambridge University Press, Cambridge.

Kant, I. (1929). *The Critique of Pure Reason* [1781-1787]. Mac-Millan, New York. Translation by N. Kemp Smith, reprint 1998.

Katsuno, H. and Mendelzon, A. (1992). On the difference between updating a knowledge base and revising it. In Gärdenfors, 1992, pages 183–203.

Konolige, K. (1992). Abduction versus closure in causal theories. *Artificial Intelligence*, 53:255–272.

Lakatos, I. (1970). Falsification and the methodology of scientific research programs. In Lakatos and Musgrave, 1970, pages 91–195.

Lakatos, I. (1971). History of science and its rational reconstructions. In Lakatos and Musgrave, 1970, pages 91–135.

Lakatos, I. (1976). *Proofs and Refutations. The Logic of Mathematical Discovery*. Cambridge University Press, Cambridge.

Lakatos, I. and Musgrave, A., editors (1970). *Criticism and the Growth of Knowledge*. Cambridge University Press, Cambridge.

Lanzola, G., Stefanelli, M., Barosi, G., and Magnani, L. (1990). NEOANEMIA: A knowledge-based system emulating diagnostic reasoning. *Computers and Biomedical Research*, 23:560–582.

Levesque, H. J. (1989). A knowledge level account of abduction. In *Proceedings IJCAI-89*, pages 1061–1067, Detroit, MI.

Levi, I. (1996). *For the Sake of the Argument. Ramsey Test Conditionals, Inductive Inference, and Nonmonotonic Reasoning*. Cambridge University Press, Cambridge.

Lloyd, J. W. (1987). *Foundations of Logic Programming*. Springer, Berlin. Second edition.

Magnani, L. (1988). Epistémologie de l'invention scientifique. *Communication & Cognition*, 21:273–291.

Magnani, L. (1992). Abductive reasoning: Philosophical and educational perspectives in medicine. In Evans and Patel, 1992, pages 21–41.

Magnani, L. (1997). *Ingegnerie della conoscenza. Introduzione alla filosofia computazionale*. Marcos y Marcos, Milan.

Magnani, L. (2001). *Abduction, Reason, and Science. Processes of Discovery and Explanation*. Kluwer Academic/Plenum Publishers, New York.

Magnani, L. (2002). Epistemic mediators and model-based discovery in science. In Magnani and Nersessian, 2002, pages 305–329.

Magnani, L. (2005). Withdrawing hypotheses using negation as failure. In Probst, S., Erdélyi, A., Moretto, A., and Chemla, K., editors, *Liberté et négation. Ceci n'est pas un festschrift pour Imre Toth*, Archive ouverte en Sciences de l'Homme et de la Société, Centre pour la Communication Scientifique Directe, CNRS, Paris.

Magnani, L., editor (2006a). *Abduction and Creative Inferences in Science. Special Issue of the Logic Journal of IGPL*.

Magnani, L., editor (2006b). *Model Based Reasoning in Science and Engineering*. College Publications, London.

Magnani, L., Civita, S., and Previde Massara, G. (1994). Visual cognition and cognitive modeling. In Cantoni, 1994, pages 229–243.

Magnani, L. and Nersessian, N. J., editors (2002). *Model-Based Reasoning. Science, Technology, Values*. Kluwer Academic/Plenum Publishers, New York.

Meheus, J. and Batens, D. (2006). A formal logic for abductive reasoning. In Magnani, 2006a.

Meheus, J., Verhoeven, L., Van Dyck, M., and Provijn, D. (2002). Ampliative adaptive logics and the foundation of logic-based approaches to abduction. In Magnani and Nersessian, 2002, pages 39–71.

Nersessian, N. (1995). Constructive modeling in creating scientific understanding. *Science & Education*, 4:203–226.

Nersessian, N., Griffith, T. W., and Goel, A. (1997). Constructive modeling in scientific discovery. Technical report, Georgia Institute of Technology, Atlanta, GA.

Nersessian, N. J. (1988). Reasoning from imagery and analogy in scientific concept formation. In Leplin, A. F. . J., editor, *PSA 1988*, pages 41–47. Philosophy of Science Association, East Lansing.

Nersessian, N. J. (1994). Opening the black box: cognitive science and history of science. Cognitive Science Report Series GIT-COGSCI 94/23, Georgia Institute of Technology, Atlanta. Partially published in *Osiris* 10, 1995, 194-211.

Pearl, J. (1988). *Probabilistic Reasoning in Intelligent Systems*. Morgan Kaufmann, San Mateo, CA.

Peirce, C. (1931-1958). *Collected Papers*. Harvard University Press, Cambridge. 8 vols., C. Hartshorne and P. Weiss, editors, vols. I-VI, & A.W. Burks, editor, vols. VII-VIII.

Peng, I. and Reggia, I. (1987a). A probabilistic causal model for diagnostic problem solving I: Integrating symbolic causal inference with numeric probabilistic inference. *IEEE Transactions on Systems, Man, and Cybernetics*, 17:146–162.

Peng, I. and Reggia, I. (1987b). A probabilistic causal model for diagnostic problem solving II: Diagnostic strategy. *IEEE Transactions on Systems, Man, and Cybernetics*, 17:395–406.

Poincaré, H. (1902). *La science et l'hypothèse*. Flammarion, Paris. English translation by W. J. G., [only initials indicated].

Poincaré, H. (1905). *La valeur de la science*. Flammarion, Paris. English translation by G. B. Halsted, *The Value of Science*, Dover Publications, New York, 1958.

Poole, D. (1988). A logical framework for default reasoning. *Artificial Intelligence*, pages 27–47.

Poole, D. (1991). Representing diagnostic knowledge for probabilistic horn abduction. In *Proceedings IJCAI-91*, pages 1129–1135, Sydney, NSW.

Pople, H. (1973). On the mechanization of abductive logic. In *Proceedings of the International Joint Conference on Artificial Intelligence*, volume 8, pages 147–152.

Popper, K. (1963). *Conjectures and Refutations. The Growth of Scientific Knowledge*. Routledge and Kegan Paul, London.

Quine, W. V. O. (1951). Two dogmas of empiricism. *Philosophical Review*, 40:113–127. Also in Quine, W. V. O., *From a Logical Point of View*, Hutchinson, London, 1953, 1961, pages 20-46.

Quine, W. V. O. (1979). *Philosophy of Logic*. Prentice-Hall, Englewood Cliffs, NJ.

Ramoni, M., Stefanelli, M., Magnani, L., and Barosi, G. (1992). An epistemological framework for medical knowledge-based systems. *IEEE Transactions on Systems, Man, and Cybernetics*, 22(6):1361–1375.

Reggia, J., Dana, S., and Pearl, Y. (1983). Expert systems based on set covering model. *International Journal on Man-Machine Studies*, 19:443–460.

Reiter, R. (1987). A theory of diagnosis from first principles. *Artificial Intelligence*, 32:57–95.

Reiter, R. and de Kleer, J. (1987). Foundations of assumption-based truth maintenance systems: Preliminary report. In *Proceedings AAAI-87*, pages 183–188, Seattle, WA.

Roussel, P. (1975). *PROLOG: Manual de référence et d'utilisation*. Group d'intelligence artificielle, Université d'Aix-Marseille, Luminy.

Shanahan, M. (1989). Prediction is deduction but explanation is abduction. In *Proceedings IJCAI-89*, pages 1140–1145, Detroit, MI.

Shepherdson, J. C. (1984). Negation as failure: A comparison of Clark's completed data base and Reiter's closed world assumption. *Journal of Logic Programming*, 1(1):51–79.

Shepherdson, J. C. (1988). Negation in logic programming. In Minker, J., editor, *Foundations of Deductive Databases*, pages 19–88. Morgan Kaufmann, Los Altos, CA.

Stefanelli, M. and Ramoni, M. (1992). Epistemological constraints on medical knowledge-based systems. In Evans and Patel, 1992, pages 3–20.

Thagard, P. (1988). *Computational Philosophy of Science*. The MIT Press, Cambridge, MA.

Thagard, P. (1992). *Conceptual Revolutions*. Princeton University Press, Princeton.

Thagard, P. and Shelley, C. P. (1997). Abductive reasoning: Logic, visual thinking, and coherence. In Dalla Chiara et al., 1997, pages 413–427.

Thagard, P. and Verbeurgt, K. (1998). Coherence as constraint satisfaction. *Cognitive Science*, 22(1):1–24.

CONCEPTUAL CHANGE: CREATIVITY, COGNITION, AND CULTURE

Nancy J. Nersessian
College of Computing
School of Interactive Computing
Georgia Institute of Technology
nancyn@cc.gatech.edu

1. Introduction

Concept formation and change—what I here call "conceptual innovation"—is one of the most creative dimensions of scientific practice. Throughout the history of the sciences changes in representational structure have provided "revolutionary" understandings of nature. As with other creative outcomes, conceptual revolutions are still widely perceived to be the outcomes of mysterious acts of individual genius, such as represented by an Isaac Newton, a Charles Darwin, or an Albert Einstein. The object of this paper is to dispel this notion by establishing how to incorporate both the undoubtedly unique contributions of individual scientists and the inherently socio-cultural nature of all scientific creations into the analysis of conceptual innovation. The route to meeting this objective lies in interpreting the conceptual practices scientists employ as deriving both from aspects of mundane human cognitive capabilities and from the social and cultural contexts, scientific and ordinary, in which they are embedded. What is required to construct such an interpretation is 1) knowledge of pertinent aspects of human cognition, 2) knowledge of specific practices implicated in cases of conceptual innovation, and 3) an understanding of how social and cultural contexts provide conceptual, analytical, and material resources that shape such practices.

2. Interpreting Conceptual Practices: Cognitive-Historical Analysis

In contemporary cognitive studies of science, the methodologies employed in investigating the practices scientists use in creating knowledge are ethnography, *in vivo* observation, laboratory experiments, and cognitive-historical analysis.

J. Meheus and T. Nickles (eds.), *Models of Discovery and Creativity*,
DOI 10.1007/978-90-481-3421-2_8, © Springer Science + Business Media B.V. 2009

Although it is possible to gain knowledge of conceptual practices by observing scientists in naturalistic settings, such as their own research laboratories, or by observing them in the setting of a problem solving experiment in the cognitive science laboratory, it is unlikely that conceptual innovation itself will be observed in these settings. It occurs infrequently and usually involves time spans longer than hours or days. Because of these facts, cognitive-historical analysis is the primary research method for investigating conceptual innovation (see, Nersessian, 1995).

Cognitive-historical analysis uses the customary range of historical records to recover how representational, methodological, and reasoning practices have been developed and used by scientists. These practices are studied over time spans of varying length, ranging from shorter spans defined by the activity itself to spans of decades or more. The records include notebooks and diaries, publications, correspondence, and material artifacts such as instruments. The historical practices are then examined in light of salient investigations of human representational and reasoning practices carried out by the fields within cognitive science. These comprise cognitive psychology, artificial intelligence, cognitive neuroscience, linguistics, cognitive sociology, and cognitive anthropology. One objective of cognitive-historical analysis is to explain the cognitive basis of the generativity of these practices. Some of the salient cognitive science research is directly on scientific cognition, but for the most part the studies are of cognition in mundane contexts. Saliency is determined by the nature of the practices under scrutiny. A "continuum hypothesis" underlies the cognitive-historical method: the cognitive practices of scientists are extensions of the practices humans employ in problem solving of a more ordinary kind within various physical and social environments. That is, human cognitive abilities give rise to and constrain scientific practices. Placing the historical practices within the broader framework of human cognitive activities aids in moving beyond the specific case study to more general conclusions about the nature and function of the scientific practices.

Margaret Boden (1990) makes a clarifying distinction between "P-creative" ideas that arise from episodes in which an individual creates something culturally available, but novel for the individual in question, and "H-creative" ideas, arising from episodes in which something fundamentally new in human history is created. Boden focused her attention on the nature of the mechanisms that lead to P-creative ideas. The primary foci of ethnographies, observations, and psychology experiments are the practices scientists use in coming to learn, appropriate, and employ existing concepts. The kind of conceptual change examined in these studies is primarily P-creative, that is, their novelty is for the individual. It is "H-creative" conceptual change I am concerned with here, that is, conceptual innovations with historical impact in that they have changed existing representations of nature. However, as Boden notes, some P-creative ideas are, of course, also H-creative. The hypothesis of continuity between mundane and scientific cognition that underlies the cognitive-historical method incorporates the possibility that mechanisms implicated in P-creative instances of conceptual innovation can also be employed in H-creative instances. For example, analogy could be a generative mechanism in both kinds of innovation (Gentner et al., 1997; Nersessian, 1984, 1992a). Thus, the findings about the conceptual practices of scientists derived from the other methodological approaches in science studies are relevant to developing a cognitive-historical analysis.

In addressing the problem of conceptual innovation, the historical practices can be investigated at the level of individuals and at the level of communities. The practices of designing and executing experiments, constructing models, using mathematical tools, devising means of communicating, and training practitioners, are all relevant to understanding the nature of conceptual innovation in science. A full analysis sets these in the social and cultural contexts of training, earlier research, knowledge base, community, collaborators, competitors, and material resources. Cognitive-historical investigations of conceptual change can focus ideographically, attempting to ferret out general cognitive factors underlying the uniquely individual dimensions of practice (see, e.g., Gooding, 1990; Nersessian, 1984, 1985, 1992a, 2002; Tweney, 1992). They can also focus on practices common to many instances with the intent to formulate a general account of how it is possible they produce the outcomes (see, e.g., Darden, 1991; Nersessian, 1992a; Thagard, 1992; Tweney, 1985). In both cases the source and generativity of such practices is located in what is, generally speaking, human. On the one hand, what is human includes those cognitive structures and capabilities humans have in common—that enable and constrain the unique application of an individual scientist. On the other hand, what is human is embeddedness in social and cultural systems. To date the focus of cognitive-historical analyses has been on the cognitive capabilities, structures and processes. Investigations of these have largely drawn from research in cognitive science within the traditional "GOFAI" ("Good Old Fashioned AI") framework.

On the traditional view, cognition comprises the representations internal to an individual mind and the processes that operate on these. Thinking is independent of the medium in which it is implemented, and the environment is represented in the content of thinking through being represented in memory. Recently, these founding assumptions of cognitive science were elaborated upon extensively by Alonso Vera and Herbert Simon (1993) in response to criticisms from within cognitive science. Following earlier work by Alan Newell and Simon (1972), the unit of analysis in studying cognition is called a "physical symbol system" (PSS). A PSS has a memory capable of storing and retaining symbols and symbol structures, and a set of information processes that form structures as a function of sensory stimuli. It makes no difference to understanding cognition whether the symbol processing is carried out by a human, a computer, or any other kind of PSS. In humans, and any natural or artificial PSS with sensory receptors and motor action, sensory stimuli produce symbol structures that cause motor actions and modify symbol structures in memory. Thus, a PSS can interact with its environment by 1) receiving sensory stimuli from it and converting these into symbol structures in memory and 2) acting upon it in ways determined by the symbol structures it produces, such as motor symbols. Perceptual and motor processes connect symbol systems with the environment and provide the semantics for the symbols. So, social and cultural environments are treated as abstract content on which cognitive processes operate. As with Simon's earlier "parable of the ant" (Simon, 1981, pp. 63–66), the complexity in human behavior is understood to arise from acting in the environment. So clearly, social and cultural factors are important to understanding cognition. However, the traditional contention is that what is important about the environment for thinking is abstracted through perception and represented in the symbols generated by the cognitive system. So, these dimensions need only be examined as residing internal to the mind of a human individual or other PSS as socio-cultural *knowledge*. One implication is that it makes little difference to understanding cognition whether the thinking is carried out in an authentic environment or in a psychological research laboratory.

The reductionism of the traditional account of cognition has led those on the social side of science studies to perceive social and cognitive accounts of science as fundamentally incompatible. Social accounts have tended to "black box" the individual entirely or to render cognitive explanatory factors inconsequential in comparison with socio-cultural factors. Indeed, the perceived *in-principle* incompatibility of cognitive and social accounts of science has led some in science studies to position themselves in opposition to cognitive analyses, as witnessed, e.g., by the now-expired "ten-year moratorium" on cognitive explanations called for by Bruno Latour and Steven Woolgar (Latour and Woolgar, 1979; Latour, 1987). As with the traditional view in cognitive science, this anticognitive stance has roots in the remnants of Cartesian dualism. The mind/body, individual/social, and internal/external dichotomies associated with Cartesianism are all in play in this stance. A "cognitive explanation" is seen as tantamount to maintaining the epistemological position that the source of knowledge is ideas internal to the mind (Latour, 1999).

What must be kept in mind in discussing scientific cognition, though, is that "thinking" is an inherently social and cultural activity. It rarely just goes on "in the head" in isolation from physical and social interactions. Even when a solitary thinker wrestles with a problem closed in her study, she is still engaged in a socio-cultural process. Educational training is present. Conversations with colleagues are recalled. Further, the process often involves external representation such as sketches and equations that are socio-cultural in origin. In science *what* one thinks about and *how* one thinks about it are highly dependent on one's socio-cultural environment. Take, for example, the quest for an electromagnetic theory by British and French scientists in the latter half of the 19th century. Their representations of the problem and their methods of analysis differed considerably. To understand how Maxwell derived the mathematical equations requires knowing that he was trained in Scotland in the methods of physical geometers and in Cambridge as a mathematical physicist; that he was located in a milieu that valued Faraday's theoretical speculations, which included teachers and colleagues such as Thomson and his penchant for analogical models; and that he was located in Victorian Britain with, among other factors, the cultural fascination with machines and mechanisms. These socio-cultural factors co-determined the nature of the theoretical, experimental, and mathematical knowledge and the methodological practices with which Maxwell formulated the problem and approached its solution. They are reflected in Maxwell's reasoning through mechanical models in deriving the equations, and one cannot understand his construction of these equations without taking these factors into account. Of course, now we re-derive them by different means. Continental physicists working on electromagnetism at the time, including the French physicist Ampère, employed quite different analytical practices and drew from fundamentally different theoretical assumptions and mathematical and physical representational structures. Differences in socio-cultural factors figure significantly into why members of these communities were not able to derive the field equations.

Clearly to produce scientific knowledge requires both sophisticated cognition and a rich socio-cultural environment. The objectives of cognitive-historical analysis are to determine what enables individual agency, while at the same time explaining how the products of individuals are communal products and how these products are transformed into communal resources, transported out of the specific localities of their construction into the accepted representational content of science. To carry this out, the difficult problem that needs to be addressed is how the cognitive, the social, and the material are fused in the processes of scientists' constructing knowledge of the world. One starting point that has significant potential for resolving the problem is reconceptualizing the notion of "cognition" along the lines of recent non-reductionist analyses that challenge traditional framing of the notion. I turn to these in the next section.

3. Cognition and Culture: Situated and Distributed Cognition

Those not engaged in or with cognitive science in the last several years continue to identify it exclusively with the "rules and representations" or "logicist" accounts of human cognition associated with "GOFAI" that initiated the "cognitive revolution". As James Greeno (1989a) points out, a framing assumption of that revolution was that "the locus of thinking" is assumed to be in an individual agent's mind, rather than in interaction between an agent and a physical or social situation" (p. 134). The founding "functionalist" assumption was that thinking or intelligence is an abstractable structure that can be implemented in many different media, independent of physical or social context. Although there are still many adherents to these assumptions, contemporary cognitive science possesses alternative accounts of reasoning, representation, and learning and richer, more contextualized studies of human cognition that have yet to be exploited by science studies. Where these accounts intersect, "cognition refers not only to universal patterns of information transformation that transpire inside individuals, but also to transformations, the forms and functions of which are shared among individuals, social institutions, and historically accumulated artifacts (tools and concepts)" (Resnick et al., 1991, p. 413).

Investigations into "situated" and "distributed" cognition focus not only on the individual but also on the social group and on the various cultural artifacts and symbol systems involved in cognitive processes. It brings these directly into the purview of research on the customary cognitive science topics of representation, problem solving, and learning. The locus of analysis is always an "activity" and the unit of analysis of an activity is a "cognitive system". Analysis of a cognitive system can focus on an individual (reconceptualized as an embodied, social, tool-using agent), a group, or the material and conceptual artifacts of the context of an activity. The goal, however, is to understand cognition as an interaction among the participants in and context of an activity as it develops over time. Much of the research in this area is not conducted in standard laboratory settings, but focuses on cognitive activities in inherently social and collective contexts: in learning situations (see, e.g., Brown et al., 1989; Greeno, 1989a; Lave, 1988), in the workplace (see, e.g., Suchman, 1987; Woods, 1997), and "in the wild" (the world at large) (see, e.g., Hutchins, 1995; Norman, 1988; Shore, 1997). Further, much of it is concerned with how meaning and understanding is created collectively and addresses, directly, the problem of how cultural representations that are variable and context-relative could have universal properties of the human mind implicated in their development.

These challengers of GOFAI argue that the traditional view has mistaken the properties of a complex, *cognitive system*, comprising both the individual and the environment, for the properties of an individual mind. Thus, the critique is aimed at the traditional analytical framework in which cognitive processes are treated separately from the contexts and activities in which cognition occurs. For example, in arguing for a distributed notion of cognition, Edwin Hutchins (1995) contends that rather than construing culture as *content*, what is required is an integrated picture in which cognition and culture are interrelated notions construed in terms of *process*. Such construal leads to a shift in theoretical outlook from regarding cognitive and socio-cultural factors as independent variables to regarding cognitive processes as inherently socio-cultural. Thus the main point of contention is not whether the environment can be accommodated, but rather, *whether accounting for environmental factors requires altering fundamental notions of the structures and processes employed in cognition*. The argument is about the very nature of cognition itself.

Broadly characterized, the challenge posed to the traditional view centers on three interrelated issues: 1) the limitation of the cognitive system to the bounds of the individual mind, 2) the nature of the processing employed in cognition, and 3) the nature of, and the need for, mental representations in cognitive processing. The literature that addresses these issues is by now quite extensive and there are significant differences within and among the perspectives, so it will not be possible to lay out the positions in detail. Rather, I will highlight features of these views that seem pertinent to interpreting scientific cognition. I begin by discussing the "situative perspective" (Greeno, 1989a) and then link aspects of the other perspectives that are salient for our purposes.

Much of the impetus for developing theories of *situated cognition* has come from studies by cognitive anthropologists and sociologists concerned with learning and with work practices. Jean Lave, for instance, has attempted to explain ethnographical studies that establish striking disparities between mathematical problem solving, competency in real-world and in school learning environments. In real-world environments, such as supermarkets (Lave, 1988), adults and children exhibit high levels of competence in solving mathematics problems that are structurally of the same kind as those they fail at solving in standard school and test formulations. Lave argues that the disparities can be explained only by construing the relation between cognition and action as an interactive process that involves essentially the resources available in a specific environment. Cognition is a relation between the individual and the situation and does not just reside "in the head". Drawing on J. J. Gibson's theory of perception (Gibson, 1979), explanations of human cognition in the situative perspective employ the notion of *attunement to constraints and affordances*. On their adaptation of Gibson's notion, an affordance is a resource in the environment that supports an activity and a constraint is a regularity in a domain that is dependent upon specific conditions.

The structure of the environment provides the constraints and affordances needed in problem solving and these cannot be captured in abstract problem representations alone. Ethnographical studies of work environments by Lucy Suchman (1987), for example, have led her to argue that contrary to the traditional cognitive science view that problem solving involves formulating in the abstract the plans and goals that will be applied in solving a problem, plans and goals develop in the context of actions and are thus emergent in the problem situation. Problem solving involves improvisation and appropriation of affordances and constraints in the environment, rather than mentally represented goals and plans specified in advance of action.

Within the situative perspective, analysis of a cognitive system can focus at different levels: on the individual (conceptualized as an embodied, social, tool-using agent), a group of agents, or the material and conceptual artifacts of the context of an activity. The goal of an analysis at any level, though, is to understand cognition as an interaction among these participants in and, the context of, an activity. Cognition, thus, is understood to comprise the interactions between agents and environment, not simply the possible representations and processes in the head of an individual. Thus situated cognition is *distributed.*

As with the situative perspective, proponents of the notion of *distributed cognition* contend that the environment provides a rich structure that supports problem solving. The focus of distributed cognition research is on the claim that an environment does not just supply "scaffolding" for mental processes, as it is viewed in the traditional perspective, but that salient parts of the environment are an integral part of the cognitive system and, thus, enter essentially into the analysis of cognition. Thus they contend that a new account of cognitive processing is required—one that incorporates what is salient in the environment in a non-reductive fashion. Salient parts of an environment are, broadly characterized, those factors that can affect the outcome of an activity, such as problem solving. These cannot be determined *a priori* but need to be judged with respect to the instance. For ship navigators, for example, the nature of the function of a specific instrument can be salient, but not usually the material from which the instrument is made. For physicists, whether one is sketching on a blackboard or white board or piece of paper is likely irrelevant, but sketching on a computer screen might be salient because the computer adds resources that can affect the outcome.

Determining the *cognitive artifacts* within a specific system is a major part of the analytical task for the distributed perspective. Various kinds of external representations are candidates. Zhang & Norman (Zhang and Norman, 1995; see also Zhang, 1997), for example, have studied problem solving with isomorphic problems to ascertain potential cognitive functions of different kinds of external representations. They found that external representations differentially facilitate and constrain reasoning processes. Specifically, they argue that diagrams are cognitive artifacts in that they do not play just a supportive role in what is essentially an internal process, but that these external representations play a direct role in cognitive processing without the mediation of an internal representation of the information provided in them. On their account, affordances and constraints in the environment are construed, literally, as memory in cognitive processing. Thus, analysis of cognition in situations of problem solving with diagrams needs to be of the cognitive system that comprises both the mental and diagrammatic representations.

Research in the situative and distributed perspectives largely consists of observational case studies employing ethnographic methods. Although these studies focus on details of particular cases and often provide "thick descriptions" of these (Geertz, 1973), their objectives differ from historical, social, and cultural studies in STS that aim mainly to ferret out the specific details of a case. Rather, the aim of the cognitive research is to understand the nature of the regularities of cognition in human activity, i.e., those aspects that are common across cases. As Hutchins has framed the objective

> There are powerful regularities to be described at the level of analysis that transcend the details of the specific domain. It is not possible to discover these regularities without understanding the details of the domain, but the regularities are not about the domain specific details, they are about the nature of cognition in human activity. (Woods, 1997, p. 171)

Currently there are many research undertakings in cognitive science that share the objective of furthering an account of cognition that construes cognition and environment in relation to one another. These include research in a wide range of areas, including the embodied nature of mental representation and cognitive development in children and animals. At present there is little or no dialog among many of these. Research in each of these areas is very much research in progress, so it tends to focus internally to an area, with not much interaction across them. It is not possible to survey the various research areas that I see as comprising a body of interconnected research in the context of this paper. Instead I will focus on one issue: how culture might shape the very nature of cognitive capacities, structures, and processes.

Comparative studies in primatology and on cognitive development have led Michael Tomasello (Tomasello, 1999; Geertz, 1973), among others, to contend that cognition is inherently cultural. He argues that culture is central to the development of uniquely human cognitive abilities, both phylogenetically and ontogenetically. He begins by posing the problem of the origins of these abilities. In terms of biological evolution, the time span is just too short to account for the vast cognitive differences that separate humans from the primates closest to us genetically, e.g., chimpanzees. From comparative studies of ontogenesis in human children and other primates, he posits that the development of the uniquely human cognitive abilities began with a small phylogenetic change in the course of biological evolution: the ability to see conspecifics as like oneself, and thus to understand the intentionality of their actions. This change has had great consequences in that the processes of imitation and innovation enabled by it allowed for the accumulation of culture through transmission—or what he calls "cultural evolution".

On the account Tomasello develops, cultural evolution is the engine of cognitive evolution. That is, the expansion of cognitive capacities in the human primate has occurred as an adaptation to culture. Significantly then, culture is not something added to accounts of cognition—culture is what makes human cognition what it is. In ontogenesis, children absorb the culture and make use of its affordances and constraints in developing perspectivally-based cognitive representations. His analysis concentrates specifically on how language development creates cognitive capacities in the processes of ontogenesis, which supports the early speculation of Lev Vygotsky (1978) (whose work has influenced the development of the situative perspective discussed above) that cognitive development is socio-cultural in that it involves the internalization of external linguistic processes. However, this does not imply that cognitive processing need be all internal or linguistic. External representations seem indispensable in complex human thinking, and their development has been central to the process of cultural transmission. Merlin Donald's (1991) analysis of the evolutionary emergence of distinctively human representational systems underscores the importance of mimesis, or re-creation such as using the body to represent an idea such as the motion of an airplane, in the developments of such external representations as painting and drawing (40K years ago), writing (6K) and phonetic alphabets (4K). Donald argues for a distributed notion of memory as a symbiosis of internal and external representation on the basis of changes in the visuo-spatial architecture of human cognition with the development of external representation. So, affordances and constraints in the environment are *ab initio* part of cognitive processing.

This research into the relations between culture and cognitive development, along with developmental research in neuroscience can be construed as moving beyond the old "nature–nurture" debate through developing an *interactionist* approach. It attempts to provide an account of how evolutionary endowment and socio-cultural context act together to shape human cognitive development. Supporting this conception, neuroscience studies of the impact of socio-cultural depravation, enrichment, and trauma in humans and in non-human primates on brain structure and processes lead to a conception of the brain as possessing significant cortical plasticity (see, e.g, Elman et al., 1998; van der Kolk et al., 1996; Shore, 1997) and as a structure whose development takes place in response to the socio-cultural environment as well as genetic factors and biological evolution.

Finally, in so connecting cognition and culture, this body of research implies that human cognition should display both species-universal cognitive abilities and culturally specific cognitive processes. Tomasello discusses some of the universal learning abilities, such as those connected with language learning, among which he includes the ability to understand communicative intentions, to use role reversal to reproduce linguistic symbols and constructions, and to use linguistic symbols to contrast and share perspectives in discourse interactions (Tomasello, 1999, pp. 161–163). Although he does not discuss these, one implication is that the cognitive processes of learning, reasoning, problem solving, representation, decision making should display culturally specific features. Recent investigations into culturally specific features of cognition by Richard Nisbett and colleagues (Nisbett et al., 2001) has implications for the hypotheses linking cultural evolution and cognitive processes. This research was inspired by the substantial body of historical scholarship that maintains that there were systematic cultural differences between ancient Greek and Chinese societies, especially concerning what they call the "sense of personal *agency*" (p. 292, italics in original). Nisbett hypothesized that these differences between "eastern" and "western" cultures, broadly characterized as holistic vs. analytic thinking (p. 293), should still be detectable in cognitive processes such as categorization, memory, covariation detection, and problem solving in contemporary cultures whose development has been influenced by ancient China (China, Japan, Korea) or by ancient Greece (Western Europe, North America). In a series of experiments with subjects in East Asian and Western cultures, and subjects whose families have changed cultural location, they examined explanations, problem solving, and argument evaluation. Some significant systematic differences were found along the five dimensions they identified in the ancient cultures: 1) focusing on continuity vs. discreteness, 2) focusing on field vs. object, 3) using relations and similarities vs. categories and rules, 4) employing dialects vs. logic and first principles in reasoning, and 5) using experience-based knowledge vs. abstract analysis in explanations.

The implications of the research of the "environmental" perspectives reviewed above for the project of an integrative account of knowledge-producing practices in science are extensive. Working them out in detail is beyond the scope of this paper. One thing is clear though: situating the problem of interpreting these practices within the framework provided by environmental perspectives on cognition affords cognitive-historical analysis the possibility of analyzing from the outset the cognitive practices of scientists as bearing the imprint of human cognitive development, the imprint of the socio-cultural histories of the specific localities in which science is practiced, and the imprint of the wider societies in which science develops.

4. Creativity in Conceptual Change: The Role of Model-Based Reasoning

As discussed in Section 2, the continuum hypothesis underlying cognitive-historical analysis holds that cognitive practices of scientists are extensions of the kinds of practices humans employ in coping with their environment and in problem solving of a more ordinary kind. The mental representations and processes used in human problem solving have developed out of an interaction between two inseparable processes: biological selection and adaptation and socio-cultural construction, selection, and adaptation. Thus, scientific cognition is shaped by the evolutionary history of the human species and by the developmental processes of the human child. Basic cognitive strategies are extended and refined in explicit and critically reflective attempts to devise methods for understanding nature. As with mundane modes of inquiry, the success of those created by science is rooted in human nature and the nature of the world.

What needs to be ascertained are the nature of the representations and of the processes employed in scientific cognition. Here I will focus on a specific kind of problem solving practice employed in conceptual innovation: "model-based" reasoning. The issue of the representational format of conceptual structures is especially significant for the problem of the nature of the reasoning through which inferences are made. Different representational formats enable different modes of reasoning. The predominant modes of analysis of conceptual change have viewed conceptual structures from the perspective of languages. Clearly concepts and conceptual structures can be represented linguistically. However, in earlier cognitive-historical analyses of conceptual change, I have proposed that from the perspective of understanding the reasoning practices leading to new concepts, conceptual structures are best viewed as models and conceptual change as a process of constructing and communicating new models. This proposal derives from extensive examination of scientific practices leading to conceptual innovation. This examination establishes, first, that conceptual innovation is a problem-solving process, and, second, that model-based reasoning practices, such as analogy, visual modeling, and thought experimenting (simulative modeling) (Nersessian, 1984, 1992a, 1992b, 1999, 1988), play a central role. My analyses draw from practices employed in physics, but investigations of other sciences establish that these practices are employed across the sciences (see, e.g. Darden, 1991; Giere, 1988, 1992; Griesemer, 1991; Griesemer and Wimsatt, 1989; Latour and Woolgar, 1979; Latour, 1987; Lynch, 1985; Lynch and Woolgar, 1990; Shelley, 1996, 1999; Thagard, 1992). Although model-based reasoning practices are ubiquitous, I am, of course, not contending they are exhaustive of the practices that generate new representational structures. I have focused on these practices because they are ubiquitous and because within philosophy these practices have not traditionally been considered significant forms of scientific reasoning, even though there is abundant historical evidence in favor of their generativity. Philosophical accounts of scientific reasoning have restricted the notion of reasoning primarily to deductive and inductive arguments. Modeling practices, when considered at all, have mainly been held to perform an ancillary role as "mere aids" to reasoning. The approach taken here is to develop a cognitive basis for these practices as productive forms of reasoning more widely applicable in human reasoning than in the specific contexts in which they are employed in science. From this basis, one can mount a case for how they are productive forms of reasoning in science and how they function in conceptual innovation.

Most of the work on representation and reasoning in the cognitive sciences comes from considering individual cognition from the traditional perspective. Here I want to place scientific cognition within the framework of the environmental perspective discussed in Section 3 by starting from the assumption that scientific cognition is always situated and often distributed. However, since individual human agents are parts of cognitive systems an accounting of their role in the cognitive processing within the system is still required. Mainstream notions of mental representation, such as concepts and mental models, and of reasoning, such as analogy, can contribute to understanding the human component, with the caveat that modification will be necessary. Thinking about such notions from the perspective of cognition as situated and distributed can aid in creating alternative formulations of these. The most radical proponents of situated cognition discount the role of mental representations in cognitive processes. However, although one might not need to invoke the notion of mental representation in explaining how people drive cars around a familiar campus or measure a quantity of cottage cheese to be eaten on a diet program, it is difficult to see how one could begin to explain complex scientific problem solving without invoking it. Much of the research in distributed cognition seems consistent with the notion of mental representation. However, what kinds of mental representations and processes to accord the individuals that constitute significant components of cognitive systems remains an outstanding research problem. This section begins to address this problem in conjunction with the hypothesis that "model-based" reasoning is central in conceptual innovation. The cognitive hypothesis of reasoning through "mental modeling" is a significant component of the case for the cognitive basis of model-based reasoning. I will try to establish that a particular notion of mental modeling is more in accord with the situated and distributed nature of scientific cognition.

4.1 Mental Modeling

The notion of a "mental model" is an explanatory construct that plays a central role in much of cognitive science. It is employed widely in theories of comprehension and reasoning. In cognitive psychology there is an ongoing controversy about the nature of human reasoning that parallels the issues raised about reasoning in philosophy. On the traditional psychological view, reasoning consists of applying a mental logic to propositional representations. Critics of this view have contended that a purely syntactical account of reasoning cannot account for significant effects of semantic information exhibited in experimental studies of reasoning (see, e.g., Johnson-Laird, 1983; Mani and Johnson-Laird, 1982; McNamara and Sternberg, 1983; Oakhill and Garnham, 1996; Perrig and Kintsch, 1985; Wason, 1960, 1968). Instead, they propose adopting a hypothesis that in many instances people reason by manipulating internal models. Advocates of the mental modeling hypothesis argue that the original capacity developed as a means of simulating possible ways of maneuvering within the physical environment. It would be highly advantageous to possess the ability to anticipate the environment and possible outcomes of actions, so it is likely that many organisms have the capacity for some form of mental modeling. Given their linguistic abilities, humans should be able to create models from both perception and description, which is borne out by the research in narrative comprehension. The centrality of mental modeling to cognition is a hypothesis under investigation in numerous domains including: reasoning about causality in physical systems (see, e.g., de Kleer and Brown, 1983); the role of representations of domain knowledge in reasoning (see, e.g., Gentner and Gentner, 1983); logical reasoning (see, e.g., Johnson-Laird, 1983); narrative comprehension (see, e.g. Perrig and Kintsch, 1985); induction (see, e.g., Holland et al., 1986); and problem solving by contemporary scientists (see, e.g., Chi et al., 1981; Clement, 1989; Griffith et al., 1996). Further, a range of empirical investigations can be garnered in support of mental models as vehicles of cultural transmission, such as those into "prototypes" in concept representation (see, e.g., Rosch and Lloyd, 1978), "idealized cognitive models" in language understanding (see, e.g., Lakoff, 1987), and mental modeling in cultural transmission (see, e.g., Shore, 1997). Because the potential range of application is so extensive, some have argued that the notion can provide a unifying framework for the study of cognition (Gilhooly, 1986). For our problem, too, the hypothesis is attractive because it opens the possibility of furnishing a unified analysis of the widespread modeling practices implicated in conceptual change.

Philip Johnson-Laird (1983) credits the philosopher, psychologist, and physiologist Kenneth Craik (1943) with introducing the notion of mental modeling. Craik hypothesized that in many instances people reason by carrying out thought experiments on internal models, where a model is a structural or functional analog to a real-world phenomenon:

> By a model we thus mean any physical or chemical system which has a similar relation-structure to that of the process it imitates. By 'relation-structure' I do not mean some obscure non-physical entity which attends the model, but the fact that it is a physical working model which works in the same way as the process it parallels, in the aspects under consideration at any moment. Thus, the model need not resemble the real object pictorially; Kelvin's tide-predictor, which consists of a number of pulleys on lever, does not resemble a tide in appearance, but it works in the same way in certain essential respects..." (Craik, 1943, pp. 51–52)

Craik maintained that just as humans create physical models, such as, physical scale models of boats and bridges, to experiment with alternatives, so too the nervous system of humans and other organisms developed a way to create mental "'small scale model[s]' of external reality" (p. 61) for simulating potential outcomes of actions in a physical environment. Mental simulation occurs by the "excitation and volley of impulses which parallel the stimuli which occasioned them..." (p. 60). This internal process of reasoning results in conclusions similar to those that "might have been reached by causing the actual physical processes to occur" (p. 51). Craik based his hypothesis on the need for organisms to be able to predict the environment, thus simulation is central to mental modeling. In constructing the hypothesis he drew on existing research in neurophysiology and speculated that the ability "to parallel or model external events" (p. 51) is fundamental to the brain.

In the first place, a mental model is a form of knowledge organization. There are two main usages of the term 'mental model' that tend to get conflated in the literature: (1) a structure stored in long-term memory (LTM) and (2) a temporary structure created in working memory (WM) during a reasoning process. The first usage focuses on how the mental representation of knowledge in a domain is organized in LTM and the role it plays in supporting understanding and reasoning. The second usage focuses on the nature of the structure employed in WM in a specific comprehension and reasoning task. In considering model-based reasoning, our analysis can be restricted to WM representations and processes. This usage maintains that mental models are created and manipulated during narrative and discourse comprehension, deductive and inductive logical reasoning, and other inferential processes such as in learning and creative reasoning. In all cases, the inferencing takes place through specific operations on the model itself. Although Philip Johnson-Laird's own research focus has been on deductive and inductive reasoning tasks, and not mental modeling in other kinds of inferencing, his 1983 book provides a general theoretical treatment of mental models as WM structures that has had a wide influence. He holds that a mental model is a structural analog of a real-world or imaginary situation, process or event that the mind constructs in WM during reasoning. A mental model is a structural analog in that it embodies a representation of salient spatial, temporal, and causal structures relating the events or entities represented. The LTM knowledge drawn upon in the activity of mental modeling need not be represented in the form of a model. Johnson-Laird's account is uncommitted on the format of the LTM representation.

Although talk of mental modeling is ubiquitous in cognitive science today, unfortunately explicit accounts of just what a specific researcher means when invoking the notion are not. There is not a single fully-developed and agreed upon hypothesis about either the representational *format* of the model, where 'format' includes *structure* and *content*, or the nature of the *processing* involved in either generating a model or reasoning by means of it. So, the notion of understanding and reasoning via mental modeling is best considered as an explanatory framework under development for studying cognitive phenomena. What the various hypotheses within the framework share is that they postulate models as organized units of mental representation on which cognitive processing is carried out in diverse activities. The preponderance of research into mental modeling is concerned with either explaining logical inferencing or specifying the knowledge contained in the models in a specific domain with respect to a reasoning task or level of expertise, and not with either the format or processing issues. Here I will try to classify the major views on the format and processing issues that can be discerned from the literature.

Preliminary to discussing the issues of format and processing with respect to mental modeling, some sorting out of the terminology used in discussing mental representation, generally, will be useful. Since its inception, there has been a deep divide in the field of cognitive science between those who hold that all representation is language-like or 'propositional' (see, e.g., Fodor, 1975; Pylyshyn, 1981) and those who hold that at least some mental representation is perceptual or 'imagistic' in format (Kosslyn, 1980, 1994; Shepard and Cooper, 1932). Herbert Simon (1977) reported that this divide "nearly torpedoed the effort of the Sloan Foundation to launch a major program of support for cognitive science" (p. 385) at the inception of the field. Even though significant clarification of the issues has taken place and considerable experimental work conducted, the issue remains unresolved and most likely will continue to be until more is known about how the brain functions.

In much of the cognitive science literature 'propositional' is often treated as co-extensive with 'symbolic', comprising language-like and perceptual representations. Here I employ the term in its narrower philosophical usage of a language-like mental encoding that possesses a vocabulary, grammar, and semantics, such as Fodor's language of thought (Fodor, 1975). A propositional representation is interpreted as referring to physical objects, structures, processes, or events descriptively. The relationship between this kind of representation and what it refers to can be evaluated as being true or false. I will use the term 'iconic', rather than 'imagistic', for different kinds of analog representations, so as not to conflate these representations with mental pictures, which are only one kind of iconic representation. Iconic representation can be highly abstract and schematic, and not picture-like at all. What differentiates an iconic representation from a propositional one is that along some dimension(s) constraints are represented in a manner that is intended as isomorphic to its real-world analog. This is how I interpret Craik's notion of a 'relation-structure'. So, for example, a mental model of the tide might only capture functional constraints as does Kelvin's real-world analog predictor. Iconic representations represent spatial, temporal, causal, and functional information in analog format and procedures for constructing and manipulating the various kinds of representations. An iconic representation is interpreted as representing objects, structures, processes, or events demonstratively. The relationship between this kind of representation and what it represents is that of "similarity".

Iconic representations are similar in aspects and degrees to what they represent, and thus can be evaluated as being accurate or inaccurate.

Because different kinds of representations enable different kinds of processing operations, propositional and iconic models support reasoning in different ways. Operations on propositional models include the customary logical and mathematical manipulations. The operations are rule-based and are truth-preserving if the symbols are interpreted in a consistent manner and the properties they refer to are stable in the environment. Additional operations can be defined in limited domains provided they are consistent with the constraints that hold in the domain. Manipulation of a model requires explicit representation of salient parameters, including structural constraints and transition states. Operations on iconic models involve transformations of the representations that change their properties and relations in ways consistent with the real-world constraints of the domain. Unlike propositional models, transformational constraints for iconic models can be implicit. For example, a person could perform simple simulative reasoning about what happens when a rod is bent without having an explicit rule, such as "given the same force a longer rod will bend farther", by employing constraints implicit in perceptual experiences.

The nature of the symbols that constitute the content of a mental model is important to processing issues. The distinction Lawrence Barsalou (1999) makes between 'amodal' and 'modal' symbols in discussing mental representation, generally, provides some clarification for thinking about mental models. Amodal symbols are arbitrary transductions from perceptual states, such as those associated with language. All propositional representations are composed of amodal symbols. Modal symbols are analog to the perceptual states from which they are extracted. Although perceptual in nature, modal symbols can be highly schematic. A cat-like image would be a modal symbol, ranging from an image of Fifi with her stripes to a more generic representation containing salient perceptual elements of 'catness' without definite feature such as stripes. The strings of letters 'cat' or 'chat' or 'Katze' are amodal symbols. Iconic representations can be composed of either. For example, a representation of the situation "the circle is to the left of the square, which is to the left of the triangle" could be composed of either modal tokens ● – ■ – ▲ or amodal tokens, standing for these entities, such as $C - S - T$. The latter is iconic in that it represents the spatial structure "to the left of" in an analog manner, but the tokens representing the entities are arbitrary.

The literature on mental models posits all possible representational flavors. Holland et. al.'s (1986) "induction" account considers mental models as propositional. On their view, mental models employ production-system type representations and are manipulated applying condition–action rules to propositional representations of a specific situation, such as making inferences about a feminist bank-teller using a model constructed of knowledge of feminists and bank-tellers. In the qualitative reasoning literature, the ontology of a mental model is represented propositionally and explicitly stated "qualitative equations" provide rules governing the possible state transitions of physical systems, such as "under condition X – move to next state" or "through behavior Y – move to next state" (see, e.g., Bobrow, 1985). The research by Johnson-Laird and colleagues (see, e.g., Johnson-Laird, 1983, 1989; Johnson-Laird and Byrne, 1993) on mental modeling in deductive and inductive reasoning tasks employs amodal iconic representations. These mental models are iconic in that they depict the salient structures among the entities in the problem, but the tokens representing entities are amodal, such as the $C - S - T$ in the example above. Making a logical inference such as *modus ponens* occurs by moving amodal tokens in a specific array that captures the salient structural dimensions of a problem, and then searching for models of counterexamples to the transformation. Modal iconic mental models—or 'perceptual models'—seem to be what Craik had in mind by an internal "'small-scale model' of external reality" (Craik, 1943). Simulation would involve mimicking physical transformations. "Depictive mental models" (Schwartz and Black, 1996b, 1996) provide a contemporary example of perceptual models. For example, in studies of gear rotation problems, Schwartz and Black argue that perceptual information is used to construct and manipulate a mental model of a set-up of machine gears. In a perceptual model, implicit knowledge embedded in physical constraints would be used to simulate possible behaviors in accord with real-world behaviors.

To aid in thinking about reasoning through simulation with perceptual models, there is an extensive literature that provides evidence that humans can perform simulative transformations in imagination which mimic physical transformations that can be recruited. The combinations and transformations using mental imagery are hypothesized to take place according to internalized constraints assimilated during perception. The literature on mental imagery indicates, for example, that people can mentally simulate combinations, such as in the classic example where subjects are asked to imagine a letter B rotated 90 degrees to the left, place an upside triangle below it and remove the connecting line. The simulation processes produce an image of a heart. Further, many experiments establish that in performing a mental simulation, such as rotating a figure, subjects exhibit latency times consistent with actually turning a mental figure around (see, e.g., Finke and Shepard, 1986; Kosslyn, 1980, 1994; Shepard and Cooper, 1932). There is also an extensive literature on spatial representation in mental models that indicates representation with respect to bodily orientation rather than a symmetrical Euclidian space (see, e.g., Franklin and Tversky, 1990; Glenberg, 1997). Additionally, research on mental modeling in discourse reasoning and comprehension tasks indicates that people can simulate various kinds of knowledge of physical situations in imaginary transformations. In these cases, too, such as when the imagined objects are separated by a wall, the spatial transformations exhibit latency times consistent with the reasoner having simulated moving an object around a wall rather than through it. Another significant line of research examines the role of causal knowledge in mental simulation. As mentioned earlier, Schwartz & Black have conducted studies focusing on gear rotations that provide evidence that people are able to perform simulative causal transformations on sets of gears, as does Mary Hegarty's research on problems with pulley systems (Hegarty, 1992; Hegarty and Just, 1989, 1994). In the gear problems, simulation ability was enhanced after the subject was told explicitly to imagine rotating the gears or given an visual display indicating simulation.

These interpretations are not without their critics from the camp which maintains that all mental representation is propositional. Zenon Pylyshyn (1981, 2001), for one, continues to argue that the data on visual mental imagery and transformation can be explained without having to invoke either the existence of imagery (as anything more than epiphenomena) or simulation. To explain the latency data, for example, he argues that the demand characteristics of the task could be such that they induce the subjects to perform calculations on how much time is required to traverse the distance and figure that into their responses. The arguments and counter-arguments on both sides of the "imagery debate" are too numerous to recount here. Again, I think the issue will continue to be unresolved for the foreseeable future, until the requisite neuroscience develops. In the meantime, there are good arguments and extensive research on mental imagery that can be recruited to develop theories of the nature of mental modeling, such as Stephen Kosslyn's (1980, 1994) theory of how transformation might take place in mental imagery.

Clearly much work remains to be done in developing a satisfactory understanding of mental modeling. What I am proposing here is that utilizing a minimalist notion provides a cognitive basis for interpreting the modeling practices exhibited in the historical records of conceptual change as indicative of mental modeling having played a central role in the past episodes. The minimalist notion is: in certain problem solving tasks humans reason by constructing an internal iconic model of the situations, events and processes that in dynamic cases can be manipulated through simulation. This will be considered more fully after we have discussed model-based reasoning. In the more mundane cases the reasoning performed in mental modeling is usually, though not necessarily, successful. For example, one usually is able to simulate successfully how to get the piece of furniture through the door, because the models and manipulative processes embody largely accurate assumptions about every-day real-world events. Admittedly it is some distance from the awkward furniture scenario and simulating causal transformations of rotating gears to employing the kinds of transformations requiring causal and other knowledge contained in a scientific theory. Further, it is likely the case that model-based reasoning does not take place all "in the head", as the furniture problem might. However, as with other kinds of representing and reasoning, it is consistent with the cognitive-historical method to consider the scientific practices as outgrowths of the mundane practice of mental modeling. In the case of science where the situations are more removed from human sensory experience and the assumptions more imbued with theory, there is less assurance that a reasoning process, even if carried out correctly, will yield "success". In the evaluation process, a major criterion for success remains the goodness of fit to the phenomena, but success in science can also include such factors as enabling the generation of a viable mathematical representation that allows for progress in spite of the lack of explicitly confirming data, such as that of Newton for gravitation and James Clerk Maxwell for the time delay in propagation for the electromagnetic field.

4.2 Model-Based Reasoning

The central problem of creativity in representational change is that of how is it possible to create something new given that the process must start with existing representations. The traditional account of reasoning as carrying out logical operations on propositional representations has been a major obstacle to understanding conceptual innovation as the outcome of reasoning processes. Because the kinds of modeling employed by scientists in discovery processes cannot be reduced to logic, they are discounted as generative reasoning. Conceptual innovation is viewed as occurring in sudden flashes of insight, with new concepts springing forth from the head of the scientist—like Athena—fully grown. This does accord with some scientists' retrospective accounts, but if one examines their deeds—their papers, diaries, letters, notebooks—these records support a quite different interpretation in most cases. As I have been arguing for some years, conceptual change results from extended problem-solving processes. The records of these processes display extensive use of practices that constitute forms of model-based reasoning: analogical, visual, and simulative modeling. Modeling practices are employed both in experimental and in theoretical settings. Embracing these modeling practices as "methods" of conceptual change in science requires expanding philosophical notions of scientific reasoning to encompass forms of creative reasoning, most of which cannot be reduced to an algorithm in application, are not always productive of solutions, and can lead to incorrect solutions.

Analyzing the conceptual innovation practices as various forms of model-based reasoning requires constructing a unified account of forms of modeling that are mostly treated separately in the literature in cognitive science, such as analogy and imagery. Although the practices of analogical and visual modeling and thought experimenting can occur separately, they most often are employed together in a problem-solving process. Examining Figure 1 exemplifies why

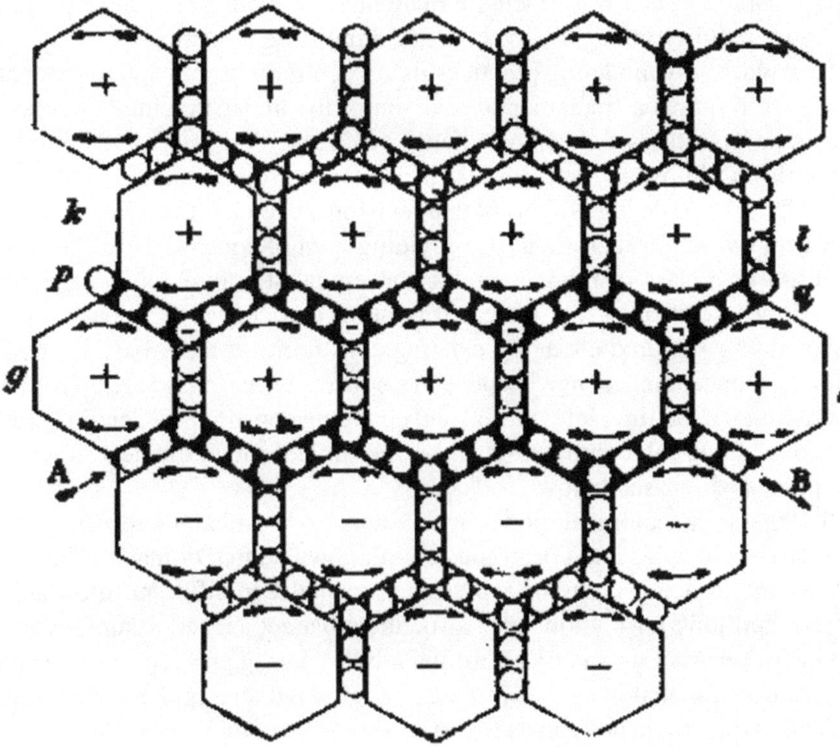

Figure 1. Maxwell's drawing of the vortex-idle wheel medium (Maxwell 1890, Vol. I, Plate VII)

a unified account is needed. The drawing is taken from a communication by Maxwell to his colleagues of his construction of a new, unified electromagnetic field concept, i.e., a paper published in *Philosophical Magazine*. The drawing is accompanied with instructions:

> Let the current from left to right commence in AB. The row of vortices *gh* above AB will be set in motion in the opposite direction to a watch... We shall suppose the row of vortices *kl* still at rest, then the layer of particles between these rows will be acted on by the row *gh* on their lower sides and will be at rest above. If they are free to move, they will rotate in the negative direction, and will at the same time move from right to left, or in the opposite direction from the current, and so form an *induced* electric current. (Maxwell, 1890, v. 1, p. 477, italics in original)

The figure is a visual representation of the analogical model Maxwell employed in constructing the electromagnetic field concept. The accompanying instructions assist the reader in animating it in thought. To understand and reason with the figure requires analogical, visual, and simulative modeling.

To explain why modeling practices figure centrally in conceptual innovation in science requires a fundamental revision of the understandings of concepts, conceptual structures, conceptual change, and reasoning customarily employed explicitly in philosophy and at least tacitly in the science studies fields more generally. The basic ingredients of that revision are to view the representation of a concept (whatever its format) as providing a set of constraints for generating members of a class of models, and a conceptual structure, as an agglomeration of these constraints. Concept formation and change is, then, a process of generating new and changing existing constraints. Model-based reasoning promotes conceptual change because these forms of reasoning are effective for abstracting, generating, integrating, and changing constraints. Genuine novelty is produced through combinations made possible through the generic abstraction process discussed below.

To engage in analogical modeling in science one calls on knowledge of the generative principles and constraints for a known source domain. These constraints and principles may be represented mentally in different informational formats and long-term knowledge structures that act as tacit assumptions employed in constructing and transforming models during problem solving. Inter- or intra-domain analogical models can be retrieved and applied with suitable adaptation, but often, and especially in cases of conceptual innovation, no direct analogy exists and construction of an initial source model is required. In these cases the analogical domain serves as the source for constraints to be used in interaction with those provided by the target problem to create an initial novel analog model, as well as subsequent models. Evaluation of the analogical modeling process is largely in terms of how well the salient constraints of a model fit the salient constraints of a target problem, with key differences playing a significant role in further model generation (Griffith et al., 1996).

As with other instances of analogical modeling, when employed in conceptual innovation the process often requires recognition of potential similarities across disparate domains, and a means of integrating information from them. "Generic abstraction" is a key reasoning process that enables recognition, adaptation, and integration. Constraints in both the target and the source domains are domain-specific. For retrieval, transfer and integration to occur in the reasoning process, they need to be understood at a sufficient level of abstraction. The various representations employed have to function with some of their features considered as unspecified, that is, as generic. In model-based reasoning processes, a central objective is to create a model that is of the *same kind* with respect to salient dimensions of the target phenomena one is trying to represent. Thus, although an instance of a model is specific, inferences made with it in a reasoning process are generic. In viewing a model generically, one takes it as representing features common to members of a class of phenomena. The relation between the generic model and a specific instantiation is similar to the type–token distinction in logic. Generality in representation is achieved by interpreting the components of the representation as referring to object, property, relation, or behavior types rather than tokens of these. In reasoning about a triangle, for instance, one cannot draw or imagine a generic triangle, but only some specific instance of a triangle. However, in considering what it has in common with all triangles, humans have the ability to imagine it as lacking specificity in the angles and the sides. That is, the reasoning context demands that the interpretation of the concrete polygon be as generic. The same is the case in considering the behavior of a physical system. To consider what a specific representation of a spring has in common with all springs, one needs to reason as though it lacked specificity in length and width and number of coils; to consider what it has in common with all simple harmonic oscillators, one needs to reason as though it lacked specificity in structure and some aspects of behavior. The analogical model, understood generically, represents what is common among the members of specific classes of physical systems, viewed with respect to a problem context.

The kind of creative reasoning employed in conceptual innovation involves not only applying generic abstractions, but creating and transforming them during the reasoning process. Maxwell's vortex-idle wheel analogy represented visually in Figure 1 is one example. In constructing the model Maxwell was considering what certain continuum mechanical and electromagnetic systems have in common. The construction and subsequent reasoning required that the dynamical relations among the idle wheels and vortices be treated as generic. That is, they must be understood to represent the class of such dynamical systems, and that class includes electric and magnetic interactions on the assumptions of Maxwell's treatment (Nersessian, 1992a, 2002). There are many significant examples of generic abstraction in conceptual innovation. In the domain of classical mechanics, for example, Newton can be interpreted as employing generic abstraction in reasoning about the commonalities among the motions of planets and projectiles in formulating a unified mathematical representation of their motions. Newton's inverse-square law of gravitation abstracts what a projectile and a planet have in common in the context of determining motion, such as that both can be represented as point masses. After Newton, the inverse-square-law model itself served as a generic model of action-at-a-distance forces for those who tried to bring all forces into the scope of Newtonian mechanics.

A variety of perceptual resources are used by scientists in modeling. Visual modeling figures prominently in conceptual change across the sciences. This may be because employing the visual modality enables the reasoner to bypass constraints inherent in current linguistic or formulaic representations of conceptual structures. As discussed in the previous section, there is a vast literature in cognitive science on mental imagery that provides evidence that humans can perform simulative transformations in imagining that mimic physical spatial and causal transformations. External visual representations provide support for the processes of constructing and reasoning with a mental model. They aid significantly in organizing cognitive activity during reasoning, such as fixing attention on the salient aspects of a model during reasoning, enabling retrieval and storage of salient information, and exhibiting salient interconnections, such as structural and causal, in appropriate co-location. Further they facilitate the construction of shared mental models in a community and the transportation of scientific models out of the local milieu of their construction.

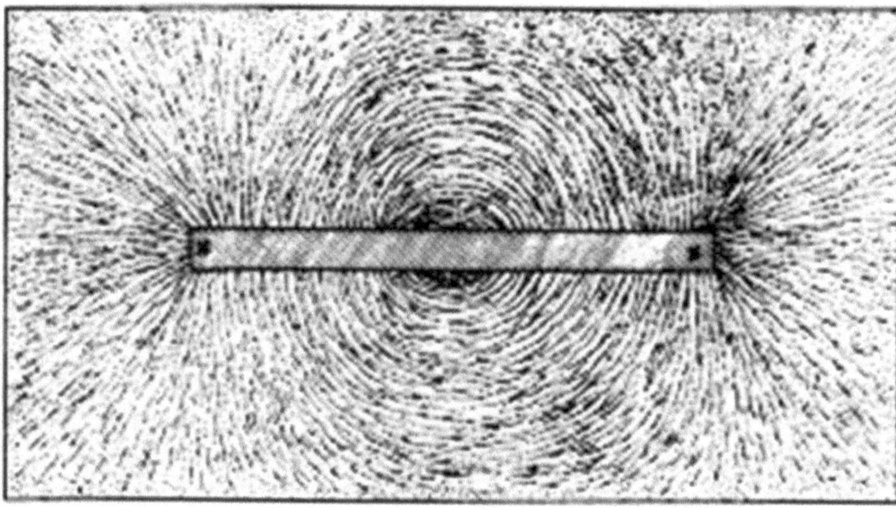

Figure 2. Lines of force

As used in model-based reasoning in physics, external visual representations tend to be schematic. These representations can aid modeling phenomena in several ways, including providing abstracted and idealized representation of aspects of phenomena and embodying aspects of theoretical models. For example, early in Michael Faraday's construction of a field concept the visual model represented in Figure 2 provided an idealized representation of the lines of force surrounding a magnet. Later in his development of the field concept, the visual model of lines of force functioned as the embodiment of a dynamical theoretical model of the transmission and interconversion of forces, generally, through stresses and strains in, and various motions of, the lines. The visual analogical model represented by Maxwell in Figure 1, however, was intended as an embodiment of an imaginary system, displaying a generic dynamical relational structure, and not as a representation of the theoretical model of electromagnetic field actions in the aether.

As a form of model-based reasoning, thought experimenting is a specific kind of simulative reasoning, which can occur in other forms of model-based reasoning. In the case of scientific thought experiments implicated in conceptual change, the main historical traces are in the form of narrative reports, constructed after the problem solving has taken place. These have often provided a significant means of effecting conceptual change within a scientific community. Accounting for the generative role of thought experimenting, thus begins with examining how these narratives support modeling processes and, by means of cognitive-historical analysis, infers that the original experiment involves a similar form of model-based reasoning (Nersessian, 1992b). What needs to be determined are: (1) how a narrative facilitates the construction of a model of an experimental situation in thought and (2) how one can reach conceptual and empirical conclusions by mentally simulating experimental processes.

From a mental modeling perspective, the function of the narrative form of presentation of a thought experiment would be to guide the reader in constructing a mental model of the situation described by it and to make inferences through simulating the events and processes depicted in it. A thought-experimental model can be construed as a form of "discourse" model (Perrig and Kintsch, 1985; Johnson-Laird, 1982), with the operations and inferences performed not on propositions but on the constructed model. Unlike a fictional narrative, however, the context of the scientific thought experiment makes the intention clear to the reader that the inferences made pertain to potential real-world situations. The narrative has already made significant abstractions, which aid in focusing attention on the salient dimensions of the model and in recognizing the situation as prototypical (generic). Thus, the experimental consequences are seen to go beyond the specific situation of the thought experiment. The thought-experimental narrative is presented in a polished form that "works", which should make it an effective means of getting comparable mental models among the members of a community of scientists. Undoubtedly experimental revision and tweaking goes on in the original reasoning and in the narrative construction, although accounts of this process are rarely presented.

Although some kinds of mental modeling may employ static representations, those derived from thought-experimental narratives are usually dynamic. The narrative delimits the specific transitions that govern what takes place. In constructing and conducting the experiment a scientist makes use of inferencing mechanisms, existing representations, and scientific and general world knowledge to make constrained transformations from one possible physical state to the next. Much of the information employed in these transformations is tacit. Thus, expertise and learning play a crucial role in the practice. So does the know-how derived from perceptual experience which David Gooding (1992) calls "embodiment". The thought-experimental process links the conceptual and the experiential dimensions of human cognitive processing. Thus, the constructed situation inherits empirical force by being abstracted both from our experiences and activities in the world and from our knowledge, conceptualizations, and assumptions of it. In this way, the data that derive from thought experimenting have empirical consequences and at the same time pinpoint the locus of the needed conceptual reform. The derived understanding forms the basis of further problem-solving efforts to construct an empirically adequate conceptualization.

All three forms of model-based reasoning are complex forms of reasoning that integrate information represented in multiple formats—propositions, models, and equations—into mental models. There are several key ingredients common to the various forms of model-based reasoning. They are semantic reasoning processes in that the models are intended as interpretations of target physical systems, processes, phenomena, or situations. The models are retrieved or constructed on the basis of potentially satisfying salient constraints of the target domain. In the modeling process, various forms of abstraction, such as limiting case, idealization, generalization, and generic modeling, are utilized. Evaluation and adaptation take place in light of structural, causal, and/or functional constraint satisfaction and enhanced understanding of the target problem obtained through the modeling process. Simulation can be used to produce new states and enable evaluation of behaviors, constraint satisfaction, and other factors.

From the perspective of conceptual innovation and change as involving processes of generating and transforming constraints, model-based reasoning is particularly effective. Analogy is a means through which constraints are abstracted from existing representations, including quite disparate domains, and integrated into models providing candidate constraints for new concepts. Thus, although analogical modeling enables arguments, the power of analogy lies in employing generic abstraction in the service of model construction, manipulation, and evaluation. Understood in this way, analogical modeling is a powerful form of reasoning, as opposed to the standard philosophical evaluation of "argument by analogy" as a weak form of reasoning. Visual modeling appears to be a highly developed and effective means of displaying constraints in a form in which humans can grasp them and follow through their consequences immediately and efficiently. As philosophers have worried for centuries, visual representations do indeed have the potential to lead a reasoner astray, but visual modeling is an effective tool for science when sufficient constraints are guiding the reasoning process. Finally, although many thought experiments can be reconstructed as arguments (Norton, 1991), their modeling function cannot be supplanted by an argument. The argument is not evident until after the thought experiment has been constructed and executed. Thought experimenting plays a crucial role in conceptual change by showing that existing systems of constraints cannot be integrated into consistent models of the physical world. Thought experimenting may facilitate recognizing the undesirable consequences of a conceptualization in much the way that experimenting by computer simulation exposes undesirable consequences of the constraints of a scientific representation. By creating a simulative model that attempts to integrate specific systems of constraints, thought experimenting enables a scientist to grasp essential points of conflict and infer their consequences more readily than they would by reasoning through the logical consequences of a representation. Once the initial experimenter understands the implications of a thought experiment, she can guide others in the community to see them as well by crafting a description of the experiment into a narrative.

This account of the generative nature of model-based reasoning in conceptual innovation lends support to the position of other contemporary philosophers (see, e.g., Cartwright, 1989; Giere, 1988) that in reasoning with or about a theory, the basic units scientists employ are not axiom systems, not propositional networks, but models. The term 'model' is used in these accounts not in the logical sense of an abstract mapping of things to terms, but in the analog sense of a structure intended as isomorphic to some aspect of a physical system. Together these hypotheses about model-based reasoning in creating and using theories make a claim that no matter how theories and concepts may *in principle* be represented, models are the mental representations with which a scientist carries out much reasoning and by means of which she thinks and understands through the lens of a conceptual structure.

5. Model-based Reasoning as Situated and Distributed Reasoning

In an obvious but non-trivial sense, model-based reasoning is situated: the scientist constructs a model and reasons in the situation it represents. Of course, here the reasoning needs to apply to the type of phenomena and not just the specific instance. On my analysis of model-based reasoning, the generic abstraction process enables reasoning in the situation to lead to inferences applying to the appropriate class of phenomena represented in the situation. The process is similar to reasoning about "triangularity" from a representation of a specific triangle. Taking the Maxwell case of conceptual innovation, in constructing the mathematical representation of the electromagnetic field concept field, Maxwell created several models of an imaginary fluid medium drawing from the source domains of continuum mechanics and of machine mechanics. On my analysis, these analogical domains served as sources for constraints used together with those provided by the target problem to create the imaginary analog models that served as the basis of his reasoning. Maxwell also employed several imagistic representations, such as that in Figure 1. In constructing the various continuum mechanical models Maxwell was explicit about creating physical situations in which to carry out the abstract reasoning involved in determining the structural relations governing electromagnetic interactions and how to represent these mathematically.

Although ignored by many philosophers and historians, Maxwell's own comments on his method of analysis are most insightful. In investigating a new area in science, Maxwell asserted that one begins with a process of "simplification and reduction of the results of previous investigation to a form in which the mind can grasp them" (Maxwell, 1855, p. 155). That process requires a "method of investigation, which allows the mind at every step to lay hold of a clear physical conception, without being committed to any theory founded on the physical science from which that conception is borrowed so that it is neither drawn aside from the subject in pursuit of analytical subtleties, nor carries beyond truth by favorite hypotheses" (ibid., p. 156) A "physical analogy" is "that partial similarity between the laws of one science and those of another which makes each of them illustrate the other" (ibid.). In Craik's terminology, Maxwell's physical analogies are "relation structures".

It does not matter whether the mechanical systems employed in the models do or do not exist in nature; all that matters is that they are "mechanically conceivable". That is, that they supply mechanisms belonging to the classes of phenomena with dynamical relational structure common to mechanics and electromagnetism. The models provide an environment in which to carry on reasoning. Throughout his reasoning processes Maxwell abstracted from the specific mechanism to find the mathematical form of that class of mechanism, i.e., of the generic dynamical structure. In this manner, Maxwell was able to formulate the laws of the electromagnetic field by abstracting from specific mechanical models the dynamical properties and relations continuum-mechanical systems, certain machine mechanisms, and electromagnetic systems have in common. In their mathematical treatment these common dynamical properties and relations were separated from the specific instantiations provided in the models through which they had been rendered concrete. The generic mechanical relationships represented by the imaginary systems of the models served as the basis from which he abstracted a mathematical structure of sufficient generality that it represented *causal processes* in the electromagnetic medium without requiring knowledge of specific *causal mechanisms*—similar to the achievement of Newton and the universal law of gravitation.

Model-based reasoning is often a distributed process, too, where the reasoning employs not only representations and processes in the head but also in the environment. Putting it paradoxically, the mental modeling process can be construed as not only taking place in the mind. When considering external representations part of the cognitive system, it is possible the process can make direct use of information in the environment. Many instances of model-based reasoning in science and engineering employ 'external' representations that are constructed for and during the reasoning process, such as diagrams, sketches, and physical models, and these can be viewed as providing constraints and affordances essential to problem solving that augment whatever the mental representations used during the process provide. One finds evidence of their use in the historical records and in current-day scientific practices. While it might be difficult to say with surety that Maxwell sketched or had a visual representation in front of him as he reasoned, there is sufficient evidence of contemporary use both in practice and in problem-solving protocols with scientists (including gestural representations). Within cognitive systems, external representations can instantiate part of the current model of the phenomena, allow manipulation, and facilitate simulative processes. The external representation can change the nature of the processing task, such as when the TIC-TAC-TOE grid is placed over the mathematical problem "15" (Zhang and Norman, 1995). Even in the simple case of simulating how to get a piece of furniture through a doorway, it is much easier to do so when the furniture and doorway are in front of you. One line of criticism against mental modeling simulation and in favor of logical reasoning over propositions has been that it is just too complex to be psychologically realistic (Rips, 1986). There are two lines along which to answer this criticism. First, not everything needs to be represented in the head to carry out a simulation. From a distributed cognition perspective, one can expand the notion of mental modeling to comprise both what are customarily held to be the internal thought of the human agent and the external representations. Simulative model-based reasoning would, under this construal, involve a process of co-constructing 'internal' models of the phenomena and 'external' models, each of which are incomplete. Understood in this way, simulating the mental model would consist of processing information both in memory and in the environment—see Greeno, 1989b for a similar view.

Second, much of the speculation about the nature of simulation comes from considering constraints of computational modeling. Psychological theories that claim simulation utilizes perceptual and motor mechanisms have the potential to provide better solutions. As discussed earlier, my analysis of model-based reasoning in conceptual change requires adopting only a *minimalist* hypothesis: that in certain problem solving tasks humans reason by constructing an internal iconic model of the situations, events and processes that in dynamic cases can be manipulated through simulation. In constructing such a model, it does however, need to be possible to utilize information in various formats, including linguistic, formulaic, and deriving from various perceptual modalities. However, the issue of whether the content of the representation is modal or amodal and what the generative processes are for creating and operating on mental models do not have to be resolved before we can make progress on an account of model-based reasoning in science. The minimalist hypothesis locates the cognitive basis for the hypothesis that the modeling practices of scientists constitute a form of reasoning through which new conceptual structures are constructed within a major thrust in the mental modeling framework. Still, I think there are some considerations weighing in favor of perceptual mental models. Developing these fully is beyond the scope of this paper, so I make only brief allusion to them in concluding this section.

As we discussed in Section 4.1 it is on the basis of extensive cognitive science research in numerous domains that mental modeling has been proposed as a fundamental form of human reasoning. It is hypothesized to have evolved as an efficient means of navigating the environment and solving problems in matters of significance to existence in the world. Following on the evolutionary hypothesis, the perceptual mental model position appears more in accord with the ability to simulate the environment. The ability should not be unique to humans, since, for example, other animals need to anticipate and predict their environment for survival purposes. In these non-human cases, perceptual and motor mechanisms would need to be employed. What makes humans unique is that we can construct models from both perceptual and descriptive information. Possibly for the human ability of logical inferencing that Johnson-Laird investigates, amodal tokens could suffice. But, what I am calling simulative model-based reasoning is closer to imaginative thinking than logical reasoning, and there is mounting evidence from neuropsychology that the perceptual system plays a significant role in imaginative reasoning (see, e.g., Kosslyn, 1994). Again, this makes sense from an evolutionary perspective. The visual cortex is one of the oldest and most highly developed regions of the brain. As Roger Shepard has put it, perceptual mechanisms "have, through evolutionary eons, deeply internalized an intuitive wisdom about the way things transform in the world. Because this wisdom is embodied in a perceptual system that antedates, by far, the emergence of language and mathematics, imagination is more akin to visualizing than to talking or to calculating to oneself" (Shepard, 1988, p. 180). He argues that although the original ability to envision by mental modeling would have developed as a way of simulating possible courses of action in the world, it is highly plausible that, as human brains have developed, this ability has been "bent to the service of creative thought" (ibid.). Once extended to scientific reasoning, for instance, the nature and richness of models one can construct and one's ability to reason would develop as one learns domain-specific content and techniques. Comparative studies of expert and novice reasoning do indicate that skill in mental modeling develops in the course of learning (Chi et al., 1981). Thus, facility with mental modeling is a combination of an individual's biology and learning. The ability of the scientist or engineer to reason about technical material through mental modeling should differ significantly from that of ordinary folk. The salient point is that it originates in the cognitive endowment of ordinary folk.

For performing simulation tasks, a mental model would need to capture the causal coherence of a system and other relevant behavioral constraints of the kinds of physical entities represented in the model and possible relations among them. These can in principle be represented in either propositional or iconic structures. However, being able to make direct use of perceptual affordances and perceptual and motor processing would increase the ease of reasoning with a mental model, and is consistent with other creatures having the ability. Unlike an amodal representation, in a modal representation perceptually-relevant information about objects, processes, situations is available directly for use. Running through a series of logical inferences such as "if I move the right corner up and to the right then the other corner will swivel to the left" or performing a set of trigonometric calculations on a model with amodal tokens for the door and the chair in proper spatial configurations, seems a more cumbersome process than simulating possible movements in a spatial configuration using a token with perceptual features approximating the chair and the door. Perceptual mental models need only be schematic in that they contain selective representations of aspects of objects, situations, and processes, making for flexibility in reasoning and comprehension tasks. Performing a simulation with a perceptual mental model is mentally re-enacting perceived information and, thus, would facilitate inferences about the real-world phenomena. The simulation should comply with the same constraints as the system it represents, such as a catcher simulating the path of a baseball and anticipating where it will land. Modal representations would have an advantage in simulation, if, as Barsalou (1999) argues, constructing a modal representation is likely to involve reactivation of patterns of neural activity in the perceptual and motor regions of the brain that were activated in the initial experience of something. This simulation ability would require that there is at least a component of long-term memory representations that is perceptually-based. On Barsalou's theory, the long-term representation of a perceptual experience—the "perceptual symbols"—can be stored separately and reactivated in thinking to create novel combinations. The major advantage with a perceptual model is that simulation would employ perceptual (and possibly motor) information and processing directly in the inferencing process. Cast in situated and distributed terms, perceptual mental models would enable a reasoner to take direct advantage of affordances and constraints inherent in the situation being modeled. The interaction with the environment could be enacted in imagination, or in a combination of imagination and external representation. The internal processing would be making direct use of the situational information in the format in which it was encoded (Yeh and Barsalou, 1996). How simulation would take place in the brain is an open question, though Kosslyn's theory of visual mental imagery, which postulates perceptual and motor processes in image transformation, might be extended to mental modeling. Shepard (1984) and others have attempted to develop a mathematical representation of psychokinetic processing in the nervous system.

6. Culture and Cognition: Implications for Creativity

We are now in a position to return briefly to the role of model-based reasoning in conceptual innovation with an eye to indicating how the analysis in the previous sections provides potential resources for explaining how cognitive, social, and material elements are fused in the representations of science. Thus far I have argued that the account needs to be rooted in the interplay between individual mental activity and the environmental context in which reasoning takes place. First, mental modeling is not just something scientists do, but is fundamental to various aspects of human existence. The claim is that in the process of developing scientific approaches to understanding nature, the cognitive tool was extended. Skill in employing it in scientific reasoning now develops through acquiring expertise. Second, scientific modeling always takes place in a material environment that includes the natural world and socio-cultural artifacts (stemming from both within science and outside of it), including instruments devised by scientific communities to probe and represent that world. Modeling employs a range of representational resources. Third, scientists employ modeling practices not only in creating representations, but in communicative attempts at creating shared understanding. That is, modeling plays a central role in creating, comprehending, transmitting, and adopting scientific representations. In short, the modeling practices and models of scientists are cognitive and socio-cultural achievements and artifacts.

In dispelling the genius mythology discussed at the outset, two points made throughout the paper need to be re-emphasized here. First of all, social and cultural context is crucial to understanding *any* creative process in science, and conceptual innovation is no exception. As discussed in the previous sections, model-based reasoning has scientists reasoning in situations. The representations and processes employed in constructing and manipulating these situations are culturally laden. Returning to the figure Maxwell drew in constructing the mathematical representation of the electromagnetic field concept (Figure 1), what it represents is the situation he created in which to carry out abstract reasoning about certain relations between electric and magnetic phenomena. It also represents his attempt to communicate his reasoning and is a representational device with the potential to create mental models similar to his own within his community. But from where did the representation in the figure derive and why did he use the "method of physical analogy" approach rather than, say, pure mathematical analysis? As was noted previously, Maxwell's educational in Scotland and Cambridge led to his training as mathematical physicist of a certain kind. This training was significantly determinative of the nature of the theoretical, experimental, and mathematical knowledge and the methodological practices with which he formulated the problem and approached its solution. The mathematical and physical representations and methods of continuum mechanics were in his tool kit, more than action-at-a-distance representations, which of course he was aware of and could use. Continental physicists working on electromagnetism at the same time employed quite different methodological practices and drew from the fundamentally different action-at-a-distance mathematical and physical representational structures. Further, the theoretical speculations of Faraday as to the active nature of space surrounding bodies and charges made continuum mechanics more salient to Maxwell in approaching the problem. Finally, William Thomson's (later, Lord Kelvin) practice of constructing mathematical representations on the basis of analogies, though different from *how* Maxwell used analogy were especially important to *that* he started from analogical models. In sum, the culture of the specific scientific environment is evident both in the representational content of the models Maxwell's constructed and in his using analogical, visual, and simulative modeling as reasoning processes at all.

Secondly, in the process of creating new concepts, concepts from all aspects of a scientist's experience are candidates for redeployment as analogical sources which, with suitable abstraction, can be applied in specific problem-solving processes. This fact helps with the problem of how wider socio-cultural context could be implicated in the context of scientific practices. Here generic abstraction, as discussed in Section 4.2 can provide a mechanism for importing representations drawn from wider culture into the representational content of science. One can, for example, interpret the historical claim that Faraday's religious views about the "unity of nature" had a significant impact on the specific form of field concept he developed (Cantor, 1985; Nersessian, 1984, 1985; Williams, 1964) in the following way. A generic concept of the unity of nature can be abstracted from the specific religious context. Redeployed with respect to the problem of the nature of physical forces, it could provide a constraint of "the unity of all forces in nature", that then facilitated Faraday's developing a dynamical model of the interaction and interconversion of all the forces of nature—chemical, electric, magnetic, gravitational—which he did by using the forms of model-based reasoning we have discussed.

To take a more challenging example, part of Maxwell's modeling process can similarly be interpreted. Maxwell's modeling processes involved adjusting multiple constraints drawn from electromagnetism, continuum mechanics, and machine mechanics. Consider Maxwell's introduction of "idle wheel particles" into his model of the electromagnetic medium in developing the field concept (Figure 1). Maxwell's first model had vortices packed in the aether without separation (illustrated by me in Figure 3). Simulating a preliminary version

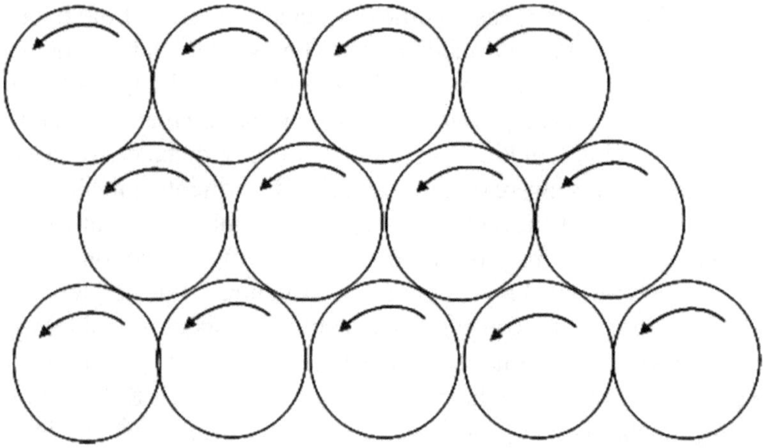

Figure 3. Cross section of model of vortex fluid medium

of the model described by Maxwell, provides a constraint that friction would bring the spinning vortices to a halt. Maxwell next utilized an intuitive model of machine gears and fly wheels. This resource is not readily connected with the hybrid of the continuum mechanical and electromagnetic domains from which he was reasoning, but it could have quickly come to mind because of a widely accessible cultural resource: the Victorian fascination with machines—especially the steam engine. Through a generic abstraction process, such as that illustrated in Figure 4, the cultural model could provide constraints to be redeployed in the vortex-idle wheel model drawn by Maxwell in Figure 1. Further model-based reasoning led to the construction of a unified mathematical representation of the electromagnetic field concept, which was the object of the problem-solving process (for a detailed technical discussion, see Nersessian, 2002).

Although these examples provide only sketches, they are based on the more detailed research cited earlier. To fully interpret any instance of conceptual innovation requires that level of analysis. A deep understanding of scientific cognition, from the mundane to the creative, and how it leads to knowledge requires no less than ascertaining what it means to be a human thinker acting in specific complex physical and socio-cultural worlds. This is a complex and multi-faceted problem. The analysis here has aimed to build a framework that will enable progress; specifically by providing a cognitive basis in support of the claim that the practices exhibited in the historical records of major conceptual innovations constitute reasoning generative of concept formation and change. Clearly more work remains to be done in filling out this account of conceptual innovation. However, the present analysis demonstrates that the perceived division between the individual and the socio-cultural, between cognition and

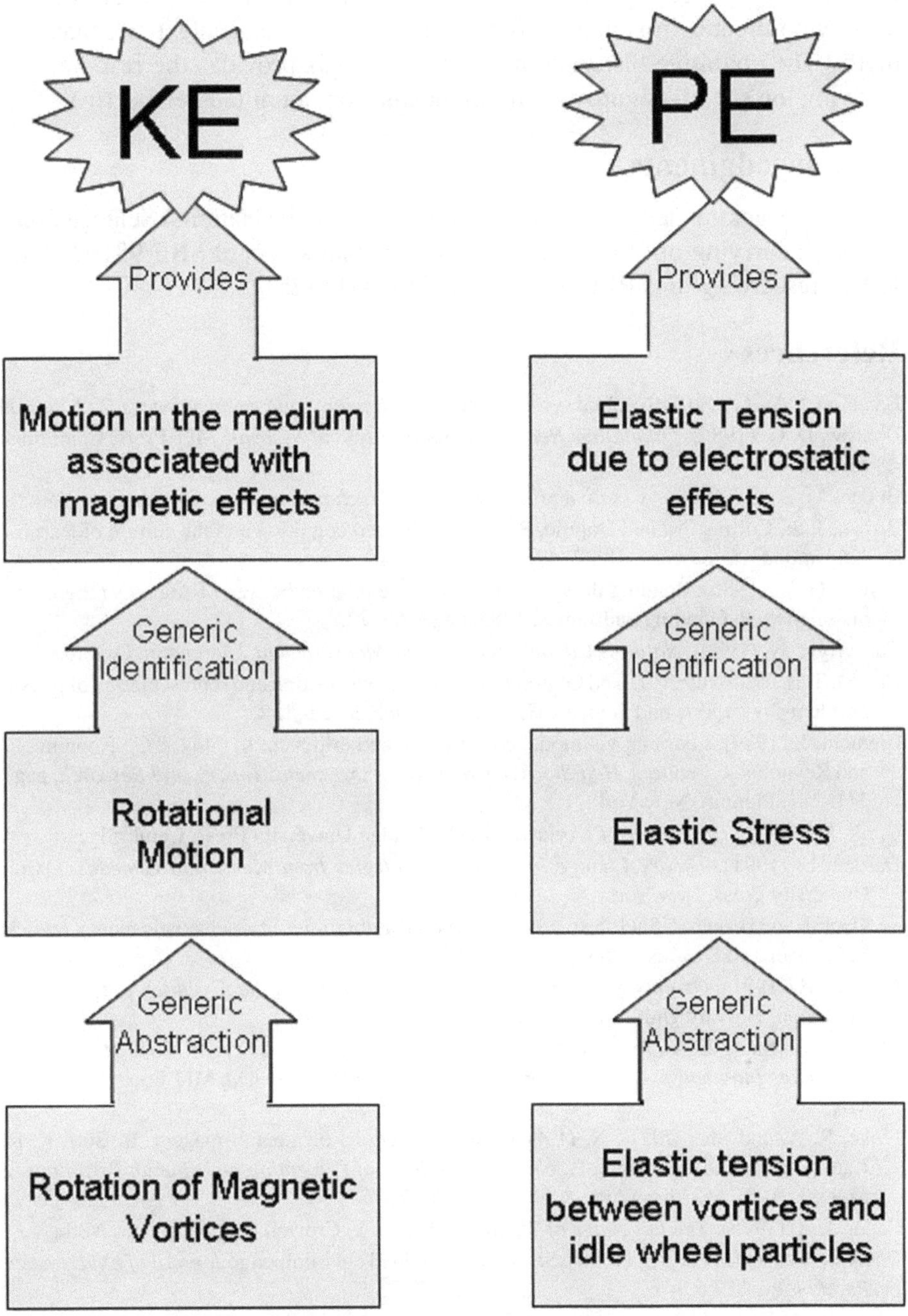

Figure 4. Identifying emergency components via generic modeling

culture, in constructing scientific knowledge is artificial. The scientific "genius" who creates in isolation from social and cultural contexts is, indeed, a myth. The cognitive-historical method of analysis provides the resources for studying the social–cognitive–cultural nexus from a unified perspective.

Acknowledgments

I gratefully acknowledge the support of grants from the National Science Foundation in carrying out this research: STS Scholar's Award SBE 9810913 and ROLE research grants REC106773 and REC0411825.

References

Barsalou, L. W. (1999). Perceptual symbol systems. *Behavioral and Brain Sciences*, 22:577–609.

Bobrow, D. G. (1985). *Qualitative Reasoning About Physical Systems*. MIT Press, Cambridge, MA.

Boden, M. A. (1990). *The Creative Mind: Myths and Mechanisms*. Basic Books, New York.

Brown, J. S., Collins, A., and Duguid, S. (1989). Situated cognition and the culture of learning. *Educational Researcher*, 18:32–42.

Cantor, G. N. (1985). Reading the book of nature: The relation between Faraday's religion and his science. In Gooding and James, 1985, pages 69–82.

Cartwright, N. (1989). *Nature's Capacities and Their Measurement*. Clarendon, Oxford.

Chi, M. T. H., Feltovich, P. J., and Glaser, R. (1981). Categorization and representation of physics problems by experts and novices. *Cognitive Science*, 5:121–152.

Clement, J. (1989). Learning via model construction and criticism. In Glover, G., Ronning, R., and Reynolds, C., editors, *Handbook of Creativity: Assessment, Theory, and Research*, pages 341–381. Plenum, New York.

Craik, K. (1943). *The Nature of Explanation*. Cambridge University Press, Cambridge.

Darden, L. (1991). *Theory Change in Science: Strategies from Mendelian Genetics*. Oxford University Press, New York.

de Kleer, J. and Brown, J. S. (1983). Assumptions and ambiguities in mechanistic mental models. In Gentner and Stevens, 1983, pages 155–190.

Donald, M. (1991). *Origins of the Modern Mind: Three Stages in the Evolution of Culture and Cognition*. Harvard University Press, Cambridge, MA.

Elman, J. L., Bates, E. A., Johnson, M., Karmiloff-Smith, A., Parisi, D., and Plunkett, K. (1998). *Rethinking Innateness: A Connectionist Perspective on Development*. MIT Press, Cambridge, MA.

Finke, R. A. and Shepard, R. N. (1986). Visual functions of mental imagery. In Boff, K. R., Kaufman, L., and Thomas, J. P., editors, *Handbook of Perception and Human Performance*, pages 37.1–37.55. John Wiley & Sons, New York.

Fodor, J. A. (1975). *The Language of Thought*. Thomas Y. Crowell Company, New York.

Franklin, N. and Tversky, B. (1990). Searching imagined environments. *Journal of Experimental Psychology*, 119:63–76.

Geertz, C. (1973). *The Interpretation of Cultures*. Basic Books, New York.

Gentner, D., Brem, S., Ferguson, R., and Wolff, P. (1997). Analogical reasoning and conceptual change: A case study of Johannes Kepler. *The Journal of the Learning Sciences*, 6:3–40.

Gentner, D. and Gentner, D. R. (1983). Flowing waters and teeming crowds: Mental models of electricity. In Gentner and Stevens, 1983, pages 99–130.

Gentner, D. and Stevens, A. L., editors (1983). *Mental Models*. Lawrence Erlbaum, Hillsdale, NJ.

Gibson, J. J. (1979). *The Ecological Approach to Visual Perception*. Houghton Mifflin, Boston.

Giere, R. N. (1988). *Explaining Science: A Cognitive Approach*. University of Chicago Press, Chicago.

Giere, R. N., editor (1992). *Cognitive Models of Science*. University of Minnesota Press, Minneapolis.

Gilhooly, K. J. (1986). Mental modeling: A framework for the study of thinking. In Bishop, J., Lochhead, J., and Perkins, D., editors, *Thinking: Progress in Research and Teaching*, pages 19–32. Lawrence Erlbaum, Hillsdale, NJ.

Glenberg, A. M. (1997). What memory is for. *Behavioral and Brain Sciences*, 20:1–19.

Gooding, D. (1990). *Experiment and the Making of Meaning: Human Agency in Scientific Observation and Experiment*. Kluwer, Dordrecht.

Gooding, D. (1992). The procedural turn: Or why did Faraday's thought experiments work? In Giere, 1992, pages 45–76.

Gooding, D. and James, F. A. J. L., editors (1985). *Faraday Rediscovered*. Stockton Press, New York.

Greeno, J. G. (1989a). A perspective on thinking. *American Psychologist*, 44:134–141.

Greeno, J. G. (1989b). Situations, mental models, and generative knowledge. In Klahr, D. and Kotovsky, K., editors, *Complex information processing*, pages 285–318. Lawrence Erlbaum, Hillsdale, NJ.

Griesemer, J. R. (1991). Must scientific diagrams be eliminable? The case of path analysis. *Biology and Philosophy*, 6:177–202.

Griesemer, J. R. and Wimsatt, W. (1989). Picturing Weismannism: A case study of conceptual evolution. In Ruse, M., editor, *What the Philosophy of Biology is: Essays for David Hull*, pages 75–137. Klwuer, Dordrecht.

Griffith, T. W., Nersessian, N., and Goel, A. (1996). The role of generic models in conceptual change. In *Proceedings of the Eighteenth Annual Conference of the Cognitive Science Society*, Hillsdale, NJ. Lawrence Erlbaum Associates.

Hegarty, M. (1992). Mental animation: Inferring motion from static diagrams of mechanical systems. *Journal of Experimental Psychology: Learning, Memory, and Cognition*, 18:1084–1102.

Hegarty, M. and Just, M. A. (1989). Understanding machines from text and diagrams. In Mandl, H. and Levin, J., editors, *Knowledge Acquisition from Text and Picture*. North Holland, Amsterdam.

Hegarty, M. and Just, M. A. (1994). Constructing mental models of machines from text and diagrams. *Journal of Memory and Language*, 32:717–742.

Holland, J. H., Holyoak, K. J., Nisbett, R. E., and Thagard, P. A. (1986). *Induction: Processes of Inference, Learning, and Discovery*. MIT Press, Cambridge, MA.

Hutchins, E. (1995). *Cognition in the Wild*. MIT Press, Cambridge, MA.

Johnson-Laird, P. N. (1982). The mental representation of the meaning of words. *Cognition*, 25:189–211.

Johnson-Laird, P. N. (1983). *Mental Models*. MIT, Cambridge, MA.

Johnson-Laird, P. N. (1989). Mental models. In Posner, M., editor, *Foundations of Cognitive Science*, pages 469–500. MIT Press, Cambridge, MA.

Johnson-Laird, P. N. and Byrne, R. (1993). Precis of the book, deduction with peer review commentaries and responses. *Brain and Behavioral Sciences*, 16:323–380.

Kosslyn, S. M. (1980). *Image and Mind*. Harvard University Press, Cambridge, MA.

Kosslyn, S. M. (1994). *Image and Brain*. MIT Press, Cambridge, MA.

Lakoff, G. (1987). *Women, Fire, and Dangerous Things: What Categories Reveal about the Mind*. University of Chicago Press, Chicago.

Latour, B. (1987). *Science in Action*. Harvard University Press, Cambridge, MA.

Latour, B. (1999). *Pandora's Hope: Essays on the Reality of Science Studies*. Harvard University Press, Cambridge, MA.

Latour, B. and Woolgar, S. (1986). *Laboratory Life: The Construction of Scientific Facts*. Princeton University Press, Princeton.

Lave, J. (1988). *Cognition in Practice: Mind, Mathematics, and Culture in Everyday Life*. Cambridge University Press, New York.

Lynch, M. (1985). *Art and Artifact in Laboratory Science: A Study of Shop Work and Shop Talk in a Research Laboratory*. Routledge and Kegan Paul, London.

Lynch, M. and Woolgar, S., editors (1990). *Representation in Scientific Practice*. MIT Press, Cambridge, MA.

Mani, K. and Johnson-Laird, P. N. (1982). The mental representation of spatial descriptions. *Memory and Cognition*, 10:181–187.

Maxwell, J. C. (1855). On Faraday's lines of force. *Scientific Papers*, 1:155–229.

Maxwell, J. C. (1890). *The Scientific Papers of James Clerk Maxwell*. Cambridge University, Cambridge, MA.

McNamara, T. P. and Sternberg, R. J. (1983). Mental models of word meaning. *Journal of Verbal Learning and Verbal Behavior*, 22:449–474.

Nersessian, N. J. (1984). *Faraday to Einstein: Constructing Meaning in Scientific Theories*. Martinus Nijhoff/Kluwer Academic Publishers, Dordrecht.

Nersessian, N. J. (1985). Faraday's field concept. In Gooding and James, 1985, pages 377–406.

Nersessian, N. J. (1988). Reasoning from imagery and analogy in scientific concept formation. In Fine, A. and Leplin, J., editors, *PSA* 1988, pages 41–47. Philosophy of Science Association, East Lansing, MI.

Nersessian, N. J. (1992a). How do scientists think? Capturing the dynamics of conceptual change in science. In Giere, 1992, pages 3–45.

Nersessian, N. J. (1992b). In the theoretician's laboratory: Thought experimenting as mental modeling. In Hull, D., Forbes, M., and Okruhlik, K., editors, *PSA* 1992, pages 291–301. Philosophy of Science Association, East Lansing, MI.

Nersessian, N. J. (1995). Opening the black box: Cognitive science and the history of science. In Thackray, A., editor, *Constructing Knowledge in the History of Science*, volume 10 of *Osiris*, pages 194–211.

Nersessian, N. J. (1999). Model-based reasoning in conceptual change. In Magnani, L., Nersessian, N. J., and Thagard, P., editors, *Model-Based Reasoning in Scientific Discovery*, pages 5–22. Kluwer/Plenum, Dordrecht.

Nersessian, N. J. (2002). Maxwell and the "method of physical analogy": Model-based reasoning, generic abstraction, and conceptual change. In Malamet, D., editor, *Reading Natural Philosophy: Essays in the History and Philosophy of Science and Mathematics*, pages 129–165. Open Court, Lacily, IL.

Newell, A. and Simon, H. A. (1972). *Human Problem Solving*. Prentice Hall, Englewood Cliffs, NJ.

Nisbett, R., Peng, K., Choi, I., and Norenzayan, A. (2001). Culture and systems of thought: Holistic v. analytic cognition. *Psychological Review*, 108:291–310.

Norman, D. A. (1988). *The Psychology of Everyday Things*. Basic Books, New York.

Norton, J. (1991). Thought experiments in Einstein's work. In Horowitz, T. and Massey, G., editors, *Thought Experiments in Science and Philosophy*, pages 129–148. Rowman and Littlefield, Savage, MD.

Oakhill, J. and Garnham, A., editors (1996). *Mental Models in Cognitive Science: Essays in honor of Philip Johnson-Laird*. Psychology Press, Philadelphia, Brighton.

Perrig, W. and Kintsch, W. (1985). Propositional and situational representations of text. *Journal of Memory and Language*, 24:503–518.

Pylyshyn, Z. (1981). The imagery debate: Analog media *vs.* tacit knowledge. *Psychological Review*, 88:16–45.

Pylyshyn, Z. (2001). Visual indexes, preconceptual objects and situated vision. *Cognition*, 80:127–158.

Resnick, L. B., Levine, J. M., and Teasley, S., editors (1991). *Perspectives on Socially Shared Cognition*. APA Press, Washington DC.

Rips, L. (1986). Mental muddles. In Brand, H. and Hernish, R., editors, *The Representation of Knowledge and Belief*, pages 258–286. University of Arizona Press, Tuscon, AZ.

Rosch, E. and Lloyd, B. (1978). *Cognition and Categorization*. Lawrence Erlbaum, Hillsdale, NJ.

Schwartz, D. L. and Black, J. B. (1996a). Analog imagery in mental model reasoning: Depictive models. *Cognitive Psychology*, 30:154–219.

Schwartz, D. L. and Black, J. B. (1996b). Shuttling between depictive models and abstract rules: Induction and fallback. *Cognitive Science*, 20:457–497.

Shelley, C. (1996). Visual abductive reasoning in archeology. *Philosophy of Science*, 63:278–301.

Shelley, C. (1999). Multiple analogies in evolutionary biology. *Studies in History and Philosophy of Biological and Biomedical Sciences*, 30:143–180.

Shepard, R. (1988). Imagination of the scientist. In Egan, K. and Nadaner, D., editors, *Imagination and the Scientist*, pages 153–185. Teachers College Press, New York.

Shepard, R. N. (1984). Ecological constraints on internal representation: Resonant kinematics of perceiving, imagining, thinking, and dreaming. *Psychological Review*, 91:417–447.

Shepard, R. N. and Cooper, L. A. (1932). *Mental Images and their Transformations*. MIT Press, Cambridge, MA.

Shore, B. (1997). *Culture in Mind: Cognition, Culture and the Problem of Meaning*. Oxford University Press, New York.

Simon, H. A. (1977). *Models of Thought*. Reidel, Dordrecht.

Simon, H. A. (1981). *The Sciences of the Artificial*. MIT Press, Cambridge, MA.

Suchman, L. A. (1987). *Plans and Situated Actions: The Problem of Human-Machine Communication*. Cambridge and New York, Cambridge University Press.

Thagard, P. (1991). *Conceptual Revolutions*. Princeton University Press, Princeton, NJ.

Tomasello, M. (1999). *The Cultural Origins of Human Cognition*. Harvard University Press, Cambridge, MA.

Tweney, R. D. (1985). Faraday's discovery of induction: A cognitive approach. In Gooding and James, 1985, pages 189–210.

Tweney, R. D. (1992). Stopping time: Faraday and the scientific creation of perceptual order. *Physis*, 29:149–164.

van der Kolk, B., McFarlane, A. C., and Weisaeth, L., editors (1996). *Traumatic Stress: The Effects of Overwhelming Experience on Mind, Body, and Society*. Guilford Press, New York.

Vera, A. and Simon, H. (1993). Situated cognition: A symbolic interpretation. *Cognitive Science*, 17:4–48.

Vygotsky, L. S. (1978). *Mind in Society: The Development of Higher Psychological Processes*. Harvard University Press, Cambridge, MA.

Wason, P. C. (1960). On the failure to eliminate hypotheses in a conceptual task. *Quarterly Journal of Experimental Psychology*, 32:109–123.

Wason, P. C. (1968). On the failure . . . — A second look. In Watson, P. C. and Johnson-Laird, P. N., editors, *Thinking and Reasoning*, pages 307–314. Cambridge University Press, Cambridge.

Williams, L. P. (1964). *Michael Faraday: A Biography*. Basic Books, New York.

Woods, D. D. (1997). Towards a theoretical base for representation design in the computer medium: Ecological perception and aiding human cognition. In Flach, J., Hancock, P., Caird, J., and Vincente, K., editors, *The Ecology of Human: Machine Systems*, pages 157–188. Lawrence Erlbaum, Hillsdale, NJ.

Yeh, W. and Barsalou, L. W. (1996). The role of situations in concept learning. In *Proceedings of the Eighteenth Annual Conference of the Cognitive Science Society*, Lawrence Erlbaum Associates, Hillsdale, NJ.

Zhang, J. (1997). The nature of external representations in problem solving. *Cognitive Science*, 21:179–217.

Zhang, J. and Norman, D. A. (1995). A representational analysis of numeration systems. *Cognition*, 57:217–295.

THE STRANGE STORY OF SCIENTIFIC METHOD

Thomas Nickles*
University of Nevada, Reno
nickles@unr.edu

> *It is more praiseworthy to produce art by deliberate design than by luck.—Aristotle*
>
> *[According to the pragmatists] there are two ways to solve a problem. You can either get what you want, or you can want what you get.—Ralph Barton Perry*
>
> *Schelling*: *But surely you do not want to give so much emphasis to the random contingencies of history!*
> *Hegel*: *Even random contingencies may yet have some sort of logic....*
>
> *Many simple ideas that seemed silly ten years ago, on the ground that they would require unthinkable computations, now seem to be valid, because fast—often parallel—computing has become commonplace.—Patrick Winston*[2]

1. Introduction

Let's begin with a story—not the full strange story of my title but an early part of it.

> One day some philosophers decided to take a walk down through history. They had passed through ancient Greece and Rome, continued on through the Middle Ages and the Renaissance, and had come to the 16th and 17th centuries, where they stumbled upon the Scientific Revolution. "What intricate design these experimental instruments and practices display!" they exclaimed. The philosophers

*This paper is an updated version of ideas that I presented at the Ghent Congress on Discovery and Creativity in spring 1998. I have tried to retain the original flavor of the paper. An ancient ancestor with the same title was presented at the University of California, Davis, and a more recent version to the British Society for Philosophy of Science. I received valuable comments at these venues. Thanks to Gaye McCollum Nickles for much help and to Joke Meheus, both for organizing the congress and for her criticisms, which I have not yet fully addressed. Thanks also to Yoichi Ishida. The paper draws upon previous work supported by U.S. National Science Foundation grants. For companion essays that provide more detail and more justification at some points, see Nickles, 2003a, 2003b. The reader should keep in mind that, today, the ideas of evolutionary computation are more familiar to, and are taken more seriously by, philosophers than in 1998.

[2] Sources of the quotes: Peirce, 1877, §1. Perry, as attributed by Reitman, 1964, p. 308. Schelling-Hegel, from a fictional interview by Solomon, 1981, p. 56. Winston, 1992, p. xxiii.

J. Meheus and T. Nickles (eds.), *Models of Discovery and Creativity*,

DOI 10.1007/978-90-481-3421-2_9,

> marveled at the intelligent order manifested in the theories and explanations. Where could all this design have come from, of a sudden? They agreed that it was virtually a priori true that there can be no design without an intelligent designer, but who or what could the intelligent designer be in this case? Since the preceding generations of inquirers had done nothing comparable, they concluded that these 17th-century natural philosophers had hit upon some intelligent *method* of discovery, a tool that amplified their intellect. For how else, short of appealing to a direct revelation from God to the innovators, could they possibly explain this explosion of successful problem-solving activity? How else could they account for the production of so much epistemically interesting design, following upon centuries of sterility?

Now this is a true[3] story, or at least an archetype with actual instances! A variant of the story is already true of Descartes, as we shall see, who, along with Bacon, is considered a founder of modern scientific method. Bacon died in 1627 and Descartes in 1650, so neither was in a position to reflect on the later work of Newton, Leibniz, and company.

In surveying the prior history of discovery, Bacon could find no rhyme or reason to it. We can construct only a chronicle of disconnected episodes of chance observation or luck. But Descartes, the mathematician, saw something different. In his survey of the previous history of inquiry and problem solving, he was struck by the mathematical prowess of the ancient Greeks; and he concluded that in order to make so many important discoveries so quickly, they must have had a method of discovery—which, however, they concealed from us. Their amazingly innovative productivity could not plausibly have been a product merely of exceptional human intellects, even with the help of luck or chance. Yet there is no design without a designer, no intelligent product without an intelligent producer, Descartes apparently reasoned. Hence, short of a direct revelation from a God they did not know, the Greek mathematicians must have had a special method of discovery. And the method must have been more than a clever tool for calculation, for it must have been conceptually powerful. It must have been analogous to assistance from a supernatural mind, able to anticipate future discovery, something capable of raising the human intellect to a nearly supernatural level. At any rate, Descartes set himself the task of rediscovering that method, or an equivalent.

For both Bacon and Descartes, method was the very antithesis of chance. No longer would the human race be a hostage to fortune. An echo of this view remained strong throughout the 19th century—that chance is the enemy of science, that attributing events to chance gives up the search for causes and deterministic laws.

---

[3]I don't mean that these history-walking philosophers had a *correct* view of history. Many scholars follow Pierre Duhem in challenging the abruptness and even the existence of the so-called Scientific Revolution.

2. Traditional Views of Method and Discovery

What is the conception of scientific method that we have received from 17th-century investigators such as Bacon, Descartes, Galileo, Huygens, Newton, and Leibniz? Scientific investigators and historical scholars ever since have been trying to figure out exactly what the methods of Bacon and Descartes (and the others) actually were. Here, in outline and whiggishly reconstructed,[4] are the salient features. For our purposes we need not worry about the fact that the idea of "science" as we now have it, and of specific sciences such as physics, chemistry, and biology, emerged only gradually from the 17th-century (or from Plato and Aristotle) to the present. Nor need we concern ourselves with the great differences among the methods proposed by the aforementioned luminaries.

It is its distinctive method that

1 demarcates science from other human endeavors.

2 accounts for the unity of the sciences as a single project, Science.

3 provides an essential definition of science. ("Science is a specific method of inquiry as well as a collection of particular results, a special process as well as the product of that process.")

4 directs scientific research (discovery) and guarantees its results (justification). (Method as the process side, or process control mechanism, of foundational epistemology.)

5 explains particular discoveries. ("Scientist S discovered d because S applied the scientific method to problem p.")

6 explains the enviable progress of science as a whole.

7 explains the Scientific Revolution[5] and rise to dominance of the modern West, but also

8 explains the rapid international diffusion of science since then.

Although early methodologists often disagreed, many of them shared something like the following conception of method, again whiggishly redescribed for our present purposes.

[4]For a justification of this sort of whiggism, see Nickles, 1992b, 1995.
[5]This is an example of a whiggish claim. It is surprising to today's generation that the idea of "the Scientific Revolution" did not become prominent until the writings of Alexander Koyré, Herbert Butterfield, and other 20th-century historians.

Some characteristics of method

1 There exists one, master scientific method as the guiding theory of the research process, although there are many efficacious local procedures—instances or applications of the general method.

2 Method is a new kind of logic or inference procedure and thus

(a) normative. Method consists of rules for productive thinking and acting in inquiry, and perhaps also a catalogue of errors to avoid.

(b) computational. Method involves mechanical, quasi-algorithmic or at least rule-based, rational, step-by-step procedures.

(c) ahistorical: Logic is not context-dependent and, in this sense, exists outside of history. (Logical possibility covers all historical possibilities.)

(d) content-neutral, a priori. Method includes no empirical claims about the universe.[6]

(e) hence, domain neutral and culturally neutral, thus universal and portable. Method

i applies to all scientific problems and fields during all historical periods.

ii transfers from one problem or field to another and from one person or research laboratory to another, regardless of nationality.

iii does not discriminate (much) on grounds of mental capacity; smoothes out differences, levels the playing field by furnishing intelligence a mental prosthetic (Bacon, Descartes), thereby minimizing this sort of constitutive luck.

3 Regarding "discovery", method satisfies 2a–2d above, plus it

(f) minimizes the need for luck, since method suffices in principle[7] to make all discoveries, including deep, postulatory theories introducing new theoretical language.

(g) contains all possible discoveries implicitly.

(h) explains discovery computationally, or at least shows how such discovery is possible.

(i) justifies the claims it produces.

---

[6]By content-neutral, I mean domain- and problem-independent. Descartes, for example, did believe in what Kant would later call synthetic a priori propositions (see below).

[7]This does not deny that some discoveries may still be made by luck.

(j) achieves item 3i in a generative (i.e., stronger than consequentialist) manner.[8]

Notice that, as originally conceived, scientific method was a method of discovery, not merely a method of justification. In fact, discovery and justification were one: the primary form of justification was that the result was discovered by use of the correct methodological procedure. The whole point of Bacon's and Descartes' provision of a method was that it would provide reliable new knowledge. Let us collect the tenets pertaining to discovery under the rubric, *the classical discovery program.*

How strange and wonderful is this scientific method—a panacea for our most pressing epistemic problems, a gift from God that gives human investigators almost supernatural powers of intellect! The method itself contains all discoveries about our universe implicitly yet in itself is free of explicit empirical content. It must apply to all possible contexts, including all possible worlds. Add the right observational information and the method produces a discovery. In effect, the method partitions inquiry into an a priori component and an empirical component.

3. Scientific Method (So-Conceived) Is Impossible

Every point on the above two lists has long been under strong attack from scientists, from philosophers, historians, and sociologists of science and from artificial intelligence experts. The attacks target the idea of a uniform logic of justification as well as that of a logic of discovery. I shall mention seven main avenues of attack and, after explaining them, devote much of the rest of the paper to one of them. The seven are

1 The problem of the criterion. A version of this ancient problem arises for any claimed "One True Method" of science. How could such a method be justified?

2 The historical rejection of logic of discovery by scientists and methodologists of science. There was a corresponding weakening of logic of justification as foundationist epistemology gave way to fallibilism.

3 The recent critique of philosophers' accounts by sociologists of science and other science studies experts.

[8]Consequential justification is based entirely on testing the logical consequences of a claim, as in the standard, hypothetico-deductive (H-D) model of inquiry. By contrast, generative justification reasons *to* the claim from prior premises. See Nickles, 1987, 1992b.

4 The still more recent "No Free Lunch" theorems from computer science and artificial intelligence (AI), which seemingly refute the possibility of a general method of science.

5 The argument from evolutionary epistemology that rejects methodology of discovery as question-begging and that returns us to the default position of chance and luck as the sources of innovation.

6 The argument from the indispensability of luck or chance for inquiry, that the possibility of luck is a necessary condition for inquiry.

7 The argument from intellectual economy against method as a form of central planning.

I shall now explain the point of each line of objection.

1. The ancient skeptics argued that the search for an absolute criterion of knowledge can never be successful, for it runs into the following dilemma. Any proposed criterion must itself be justified. It is either self-justifying, which is viciously circular, or justified by some deeper criterion, which begins a vicious logical regress. A similar problem faces claims to have found "the One True Method" of science.

Bacon's and Descartes' remarkably strong claims for method raise the question of how exactly they could have found and validated their methods, for they themselves insisted that no reliable method of inquiry was available to them. Their answer is consistent with their views of previous history: they hit upon their methods by luck. (E.g., in *Discourse on Method* of 1637, Descartes tells us quite explicitly that he arrived at his method by the good fortune of his life and course of studies.) As each founder would acknowledge, finding "the method" was the luck to end all luck in scientific inquiry. The big question remaining is how each man could be so sure that his method would work nearly infallibly. For, in terms of the distinction introduced above (and to be more fully explained below), Bacon's and Descartes' justification of their methods could be neither generative nor consequential. It could not be justified by the method of its own generation, for no reliable method was previously available, only luck; and the question would then arise all over again as to what justifies the prior method. But neither could it be justified in terms of its consequences, for few were yet available. It would be two centuries before William Whewell could look back on the many scientific achievements of the preceding centuries and attempt to justify a method on this basis—and by then a quite different method from those of Bacon and Descartes! Well, then, can we say that method can be retrospectively *self*-justifying? Perhaps it can bootstrap itself into respectability in this fashion, but, again, such a justification was not available to the 17th-century founders.

Descartes seems to have considered finding his method as somewhat analogous to the discovery of a demonstration of geometry. The path to the final result does not really matter, since the final result is somehow self-certifying. But this position only makes more urgent the question how such a method, certified in advance of any empirical inquiry, can somehow implicitly contain all future, and presumably all possible, discoveries.[9] Furthermore, a proof is a proof only against a criteriological background of rules—a prior method—and, again, none was available to the first methodologists other than the very syllogistic logic that they claimed to transcend. (Besides, what justifies the basic principles of logic or geometry? Are they self-evident to reason?) Both Bacon and Descartes, in very different ways, seem committed to what Kant would later call synthetic a priori knowledge: something fundamental was just *given* to us, without the need for inquiry but capable of furnishing a basis for inquiry into the rational structure of the universe. Hardly anyone today would defend such a position. To us the classical discovery program seems utterly unjustified.

2. In his well-known article, "Why Was the logic of discovery Abandoned?", Larry Laudan (1980) insightfully outlined the history of late 18th- and early 19th-century methodology of science, during which time scientists and methodologists largely abandoned Baconian and any remaining Cartesian methods in favor of self-corrective methods that frankly acknowledged the fallibility of scientific claims and practices. Specifically, the method of hypothesis gradually replaced Baconian-inductive and Cartesian-Newtonian deductive methods. Newton had famously refused to "feign hypotheses". To be sure, he did employ hypotheses in his research; but in both the *Principia* and the *Opticks*, he cast his final results in the form of geometrical proof. In this form it was supposedly evident how those results could have been produced by an idealized discovery procedure.[10] But once we adopt a thoroughly fallibilist attitude, the epistemological situation changes completely, for now there is no chance that a hypothesis can ever graduate into a fully proven theory.

Furthermore, working forward step by step from previously observed facts, in accordance with the inductive method commonly attributed to Bacon, was too restrictive. Since the early 17th century, most natural philosophers had been committed to the view that the world of common experience bears scant resemblance to underlying reality. They had maintained a sharp appearance-reality

[9]Descartes himself realized that empirical inquiry was necessary beyond a certain point, since his a priori principles did not determine which of various alternative mechanisms God may have chosen by his own arbitrary will for specific phenomena.

[10]See Nickles, 1984, 1985 on the common confusion of original discovery with the reconstructed discovery path that I term 'discoverability' or 'generatability' or 'generative justification'. (See also Notes 14 and 24.) In the main text I say 'seemingly' because even an original proof can only be discovered by trial and error (see §6).

distinction. Baconian methods had little chance of disclosing the underlying, explanatory causes of the observed facts. Meanwhile, progress was being made on several fronts by people who did risk hypotheses and then tested them against experience, an exercise that could flourish even in the absence of robustly Baconian data sets.

For these and other reasons,[11] methodologists gradually found their way to the view most explicitly advocated by Popper (1934, 1963, 1972) and some of the logical empiricists in the 20th century: it does not matter how we arrive at our hypotheses or other problem solutions, only how we test them. It is only the testable consequences, not the logical antecedents, that count. Hence, justification no longer depends upon the method of discovery, even if there is one. Laudan usefully dubbed this position *consequentialism* as opposed to the earlier *generativism*.

3. Most science studies experts agree that the existence of a general scientific method is a myth that has been taken far too seriously, even by those philosophers who have critically examined it. Indeed, philosophers, given their professional biases, have naively taken at face value the methodological claims made by scientists down through history, when genuine historical research shows these, in nearly every case, to be nothing more than *post hoc*, rhetorical overlay on the work actually accomplished. Scientific inquiry, the sociologists rightly point out, can be systematic in its practices without following a four-step method. In short, they, too, reject the story with which we began. They replace that story with no uniform account, but they normally stress the utter contingency of historical developments of all kinds, including scientific developments.

For example, social historian John Schuster (1977) has criticized, nay ridiculed, Descartes' methodological pretensions. There is no "single, transferable method responsible for the progress of scientific knowledge" (Schuster and Yeo, 1986, p. ix). Schuster has a point. I would add that the traditional story of scientific method simply parallels the Biblical master narrative, with copious method standing for the Tree of Knowledge from which we may now eat (Old Testament) or as the surrogate for Jesus Christ (New Testament), the portable (ecumenical) message of hope, the straight-and-narrow way to overcome our epistemic fall from grace.[12] Alternatively, the idea One True Method as universal, all-powerful, and beneficent resonates with monotheism. This should not surprise us, for the early investigators had to construct their positions out of

[11] See Laudan, 1981 and also Nickles, 1987.

[12] Recall that Descartes' intellectual autobiography is the story of his descent into the epistemic hell of total skepticism and of his victory over skepticism as imaginatively embodied in the Evil Demon.

the cultural resources available to them, and those resources were dominated by Christian theology.

Historians and sociologists of science usually prefer to speak of social constructions rather than discoveries, since the latter word implies a commitment to strong epistemological realism—as if scientists have succeeded in uncovering something just waiting there, *so described*, to be discovered. Clearly, the founders of modern methodology were committed to such a view. I am not. In what follows I shall continue to use the term 'discovery', since it identifies an established topic area, but in a non-doctrinaire sense that does not presume the truth of what is discovered. For convenience, I shall dub any significant innovation a "discovery".[13]

4. The relatively short history of AI recapitulates, to an interesting degree, the modern history of science in its movement from general, content-free methods to domain-specific, knowledge-based programs and beyond. From about 1956, founders Herbert Simon and Allen Newell conceived intelligent problem-solving programs as logical inference systems reasoning from relatively few general heuristic rules such as hill climbing and backward chaining. They regarded these rules as roughly analogous, in the problem-solving universe, to Newton's laws. Such was their Logic Theorist and the more notable General Problem Solver (GPS).[14] When that approach failed to produce powerful gen-

[13]Historians and sociologists have further shown that even major discoveries can be difficult to characterize. Thomas Kuhn (1962, §1) famously discussed the problem of determining who should be credited with the discovery of oxygen and what exactly it was that that person (or those people) discovered. Kuhn (1978) repeated the exercise in far more detail for the question, Who discovered the quantum theory? More recently, Robert Olby (1979), Augustine Brannigan (1981), and Simon Schaffer (1986, 1994), to take three of the most prominent examples, have shown how difficult are the problems of recognizing and crediting innovations, both for the relevant scientific communities themselves and for historians, philosophers, and sociologists of science using historical materials. To be a discovery, the corresponding claims and practices have to be legitimated by the relevant specialist community, a complex process that typically involves negotiation. Moreover, the attribution of discovery X to person P serves very different functions within the scientific communities and their lay audiences than getting the history right. Besides, in most cases it is a mistake to attribute "the" discovery to an individual. In the case of deep discoveries, it usually takes years, even decades, to refine, interpret, and reinterpret the significance of the original results, a multi-pass process that typically involves numerous people doing various kinds of research and receiving the necessary support and recognition from institutions such as funding agencies, conference organizers, and journal editors. No one working at the frontier of research can possibly appreciate the full implications of their work.

I agree with this complex conception of "discovery". However, my essay does not pretend to address discovery in its full social trappings. Accordingly, I shall simply assume that the work I describe or imagine is embedded in appropriate specialist communities and supporting cultures. I shall focus on "method of discovery" in the sense of usable problem-solving routines and practices and whether or not they can be innovative and reasonably general in scope. Perhaps a way to put the question is to ask whether there could be a method or a computer program that could function, as it were, as a bright, creative colleague within a research community.

[14]See Newell and Simon, 1972. Simon's later BACON series of programs, which claimed to rediscover Kepler's laws, Black's law, and others from given data sets, were more complex but ran into other problems as well (see Langley et al., 1987). As many writers have pointed out, these programs were given relatively

eral problem solvers, the AI community went to the opposite extreme with knowledge-based computation, in which a problem-solver embodies great deal of domain-specific knowledge. Accordingly, knowledge-based systems are extremely specialized. One can hardly expect a system designed to do medical diagnosis to play chess at all, even badly. This approach met with considerably greater success but, in the end, it, too, has been rather disappointing, for many of the same reasons why it has been so difficult to formulate problem-solving methods in particular scientific domains (Nickles, 2003b), but especially as a source of innovation. For even where success has been notable, the knowledge-based systems typically solve only routine problems. Even if and when the problem of transferring human problem-solving expertise to a machine is overcome, we still end up with that—the same expertise now in a machine. But that is no more knowledge-expanding than teaching a human student, and often less. Thus such systems fail to address the primary methodological problem of new knowledge that Bacon and Descartes set out to solve. The latter could raise the same "nonampliative" complaint against such systems as they in fact raised against syllogistic logic.

More recent methods, including case-based reasoning, model-based reasoning, connectionism, and evolutionary computing may prove more productive, at least in certain domains. I shall return to evolutionary computing below.

What I want to call attention to here are the recent series of theorems, the so-called "No Free Lunch" or NFL theorems, proved by David Wolpert and William Macready (e.g., Wolpert, 1996; Wolpert and Macready, 1997). The theorems purport to show that no method can be justified a priori, that no method dominates any others when averaged over all possible worlds. The NFL theorems therefore claim that there can be no universal method that works (let alone that works best or even well) in all possible worlds.

5. Evolutionary epistemologists such as Popper (1972) and Donald Campbell (1974a, 1974b, 1997) claim that trial and error underlies all innovation and that this fact drives the final nail into the coffin of method of discovery. Their position will be the focus of this essay. Their two central theses are:

> Thesis 1. An evolutionary process of blind variation plus selective retention (BV+SR) underlies all learning and innovation.
>
> Thesis 2. Thesis 1 implies the impossibility of a *method* of learning, innovation, or discovery.

---

clean problem situations and only slightly noisy data sets compared to the horribly messy situations that Kepler, Black, et al. had to face. I claim that Simon's programs model not original discovery but what I have called *discoverability* or *generatability*—the final, cleaned up version of an idealized discovery argument that may be used to justify the final conclusions. These are the "Baconian" analogues to Newton's final results, laid out *in more geometrico*. For details, see Nickles, 1984, 1985. See also Notes 10 and 24. For a history of AI, see Crevier, 1993.

A broad scattering of methodologists, epistemologists, and psychologists—from Darwin himself, William James, and Paul Souriau in the late 19th century to Popper, Campbell, Richard Dawkins, David Hull, Daniel Dennett, and Henry Plotkin in our time—have defended versions of the first thesis.[15] Popper and Campbell, among some others, have gone on to embrace Thesis 2.

Although he regarded "the problem of the growth of knowledge" as the central problem of philosophy, Popper vigorously defended his version of the hypothetico-deductive (H-D) method, his "method of conjectures and refutations", that explicitly excludes the possibility of a logic or method of discovery. For Popper, science (and any sort of knowledge-seeking or innovative enterprise) evolves in a quasi-Darwinian manner. And Campbell, the person who developed and defended Thesis 1 most fully, insisted that once investigators have applied the leverage of any available empirical, theoretical, and methodological constraints and heuristics, at the frontier of research, then they can only proceed blindly. In this respect innovative inquiry depends crucially upon a naturalistic selection process. We possess no a priori, rational faculty by means of which we can intuit the basic structure of the universe, compose a path-breaking symphony, or construct a new, more efficient fuel cell.[16] At a certain point we can only proceed by blind groping—by poking in the dark and checking the consequences.

Campbell, Popper, Dawkins, Dennett, and others (and, from a different quarter, various science studies experts[17]) argue quite persuasively that any other account presupposes the existence of supernatural powers of prescience or precognition. Taking a thoroughly naturalistic approach to human cognition, these authors make two main points. First, there is no evidence for supernatural faculties, either in humans at large or in those people that society has dubbed geniuses. This would be a most unscientific way of explaining scientific success! Second, even if such powers existed, they would not ultimately answer the question of the *sources* of innovation, for their postulation just regressively postpones the answer. If Einstein got his ideas by direct inspiration from God or from a special innate ability to know the universe, then his own originality is severely compromised. To some degree, a similar point holds for the use of a universal method that somehow already implicitly contains future discoveries. In the latter case, the inquirer becomes a kind of Socratic inquirer, prying the secrets from the method by presenting it with various problems and bits of evidence. For the method, just like the slave boy in Plato's dialogue,

[15]See Popper, 1972; Campbell, 1974a, 1974b, 1997. Campbell (1974a) provides a large bibliography of work to that point.

[16]Again, when I employ the philosophically conventional term 'discovery', I use it in a very broad sense. For present purposes, I draw no sharp distinction among discovery, invention, and construction.

[17]For an opening shot from the Edinburgh Strong Programme in Sociology of Scientific Knowledge, see Bloor, 1976. Andy Pickering (1984a, 1984b) against "the scientists' account" is highly relevant.

Meno, contains the knowledge of the universe "innately". The problem is to bring this suppressed knowledge explicitly to human consciousness. Indeed, we can perhaps better characterize this view by saying that the system consisting of a human investigator *plus* the method is highly analogous to Meno's slave. A good scientific community then consists of individuals who are good self-questioners, as Socrates himself presumably was.[18]

If all these critics of methodology of discovery are even remotely correct, then, to some significant degree, innovation depends upon luck or chance. This, of course, is anathema to traditional methodologists as well as to early AI experts, and it apparently implies Thesis 2. For the idea that there could be a method of innovation based upon luck or chance or serendipity looks positively oxymoronic. Chance and luck are the very things that method traditionally is supposed to exclude. For example, Campbell (1974, p. 428f) quotes Souriau approvingly and at length as he defends his conclusion, already in 1881, that "le principe de l'invention est le hazard".

6. If luck is unavoidable in inquiry, if inquiry presupposes luck, then the classical discovery program is doomed from the start. Here is an argument that is in fact the case.

1 Ignorance is both necessary and sufficient for luck. (Rescher, 1990)

   (a) Luck presupposes ignorance. (Limiting case: God can't get lucky.)

   (b) Wherever there is ignorance, lucky outcomes are possible.[19]

2 Thus, luck is epistemic in the sense that it depends on the state of our knowledge.

3 Now inquiry also presupposes ignorance.

4 Therefore, no method or program of genuine inquiry can rule out luck (by 1b and 3).

5 Thus, insofar as a proposed method does rule out luck completely, it also makes inquiry impossible. Such a method would have to be omniscient.

6 Thus 'omniscient method of *inquiry*' is a contradiction-in-terms.[20]

7 And so is 'luck-free method of inquiry'.

---

[18] Yet—worrisome thought!—Socrates himself always denied that he had the answers!

[19] I assume also a normal background of goals that are desired. Reformulating the argument in terms of chance rather than luck would make this assumption unnecessary.

[20] Anyone believing in an omniscient method of inquiry impales himself on the first horn of the Meno paradox: You cannot genuinely inquire if you already know the answer. For my use of the paradox, see Nickles, 2003a, 2003b.

So we have the dilemma that method is either compatible with luck (chance) or not; and, either way, method is impossible.

If we wish to challenge the apparently impossible, while accepting the above argument, then we see that any method capable of generating interesting, new knowledge must incorporate an element of luck, chance, or contingency. Popper would agree that luck is unavoidable in achieving the growth of knowledge: by nonrational means we must find a hypothesis that might solve our problem, then we must test this conjecture against nature. Here luck is involved in at least two places. However, Popper adamantly insisted that luck was not part of his method proper, since method is purely deductive.[21] So while his overall conception of science (his methodology) requires luck for progress, his method proper does not possess the resources to handle this "requirement".

If we grant that the classical discovery program cannot survive in its strong form, the question whether there can be a method consistent with luck becomes urgent. Popper and Campbell, in effect, provide a premise to add to the above argument:

8 The BV+SR model is the only defensible model of inquiry.

9 Thus innovative inquiry is not merely consistent with luck but must positively incorporate it!

In this sense, any method of inquiry would have to be serendipitous at its very core. But, again, this is precisely what pushes them to the opposite position—that there can be no such method. The idea of a goal-directed, systematic method requiring serendipity for its success certainly sounds strange. More than strange, for the phrases 'serendipitous method', 'methodological serendipity', and 'methodological luck' appear to be contradictions-in-terms. They certainly were for the classical conception of method—and for Popper and Campbell, too.[22]

In 1976 Bernard Williams and Thomas Nagel opened a vigorous debate over whether there could be such a thing as *moral* luck. As Williams pointed out, the very idea of ethics among the ancient Stoics and Epicureans was a set of rules or form of life that would make oneself an autonomous agent, morally immune to the contingencies of life, the vagaries and vicissitudes of history. The very point of ethics, on this philosophy of life, is to neutralize luck. However, both Williams and Nagel themselves, in different ways, ended up defending the possibility, nay the actual existence, of moral luck.

---

[21] This, however, is a *non sequitur*, since innovative inquiry normally requires luck even in purely deductive branches of logic and mathematics. What counts is not the abstract *existence* of deductive relations but the fact that we are ignorant of them and must search for them.

[22] Since Popper and Campbell agree, the availability of an alternative conception of method would show them to retain a whiff of the classical conception.

Be that as it may, the idea of *methodological* luck looks more problematic. One might be lucky as a scientist, as Wilhelm Roentgen, Alexander Fleming, and "Lucky Jim" Watson were, but what could it mean to be *methodologically* lucky, or for method itself to incorporate luck?

The foregoing gives a new twist to an old, romantic objection, that no *method* could ever be truly creative—by definition. For a necessary condition for something to be genuinely innovative is that it not be producible by an already available, routine practice. It is supposedly common sense that routine processes can produce only routine products (the basis of lady Lovelace's claim that computers cannot be creative). Romantics, including Popper, have always held that the springs of creativity reside in some kind of uncontrollable, spontaneous inspiration.

7. The argument from intellectual economy against a normative method as a form of central planning maintains that a rigidly enforced method would stifle scientific intellectual creativity just as central planning stifles economic development. In both cases the centralization creates an informational and computational bottleneck.

The traditional idea of method presupposes that a process governed by central rational planning is the most powerful and most efficient. For anything less will involve elements of luck, that is, blind chance. On this view, the economy of research should be a rationally planned economy, with method itself functioning as a sort of rationality czar, central intelligence agency, or (less colorfully), centralized control structure. In essential respects this model is pre-Darwinian. The model reflects the usual hubris of thinking that human reason, whatever it is, is the most powerful problem solver available to us. Like the creationist arguments against evolution, it supposes that intelligence comes first, that there is no intelligent design without an intelligent designer. The epistemic corollary would seem to be that you can't get more knowledge from less, more epistemic design from less. This is a kind of epistemic conservation principle (see below), and its plausibility may be one reason why traditional methodologists did not wonder at the fact that they could have this super-intelligent method of science before they had even made many significant discoveries about what the world is like.

There are actually two classical control theories, and Descartes formulated early versions of both. One is the linear, chain-of-causes model later improved by Newtonian mechanics: one event causes an effect, which in turn becomes at least a partial cause of another event, and so on. By controlling the causal inputs to a system in the right sort of way, we can control the output. (This much was strongly anticipated by Bacon.) The other control system advocated by Descartes was based on logical control. Here the relations between items are not causal but logical, anticipating the logical or informational control systems of today. But Descartes' logical control system was also linear, foundational, and cumulative or additive and did not involve feedback in a principled way. The idea was, rather, that if we do things strictly correctly from the beginning, that is, in correct methodical order, then we can keep adding to the foundations already laid. Since errors will never arise, we need never return to correct what went before.

Darwinian evolution amounts to the discovery (anticipated by Hegel and others) of a control theory quite different from either of the classical systems, a point to which we shall return in §5.

4. Reasons for Optimism?

The conclusion from §3 would seem to be that the classical discovery program is dead. The reasons seem overwhelming, since many lines of argument lead to roughly this same conclusion. The idea of a *method* or useful process of discovery or problem solving seems impossible. After all, how could there be a method of getting from what we know to something more—what we don't know? How can we expect to get more from less? Getting something for nothing would seem to be exactly the free lunch that we are not entitled to have.

Before giving up, however, let us look more carefully at a possibility rejected above. Consider biological nature. We now believe that the incredible variety of adaptive design that we find among the flora and fauna is the product of an evolutionary process. Five points are worth keeping firmly in mind.

1 Biological evolution is the most creative process that we know. What is more creative than biological evolution?

2 The process creates more design from less, not in the sense of striving to reach some goal but in the sense of ramifying increases in complexity as species combine with the changing environment to generate niches within niches within niches. In this special sense of getting more from less, which is one sort of something from nothing (non-conservation), biological evolution is an existence proof of creation *ex nihilo*!

3 As Darwin already realized and as subsequent research and observation have amply confirmed, far from being an almost impossible process, evolution is virtually inevitable when the essential mechanisms are in place and the environment changes relatively slowly. It can't be stopped.

4 Although biological evolution takes myriad forms, there is a skeletal process at its base, again the one already sketched by Darwin: a mechanism of variation combines with a mechanism of selection and a mechanism of

retention to produce evolution.[23] Even Stephen Jay Gould and Richard Lewontin (1979) and other critics of the adaptationist paradigm agree that there is enough method to the madness of biological evolution to make seeking a general theory worthwhile.

5 Biological evolution can be interpreted, metaphorically, as an innovative problem-solving process. For example, it has solved the problems of teledetection, that is, detection-at-a-distance, many times over by means of many kinds of vision, hearing, smell, sensitivity to vibration, and so on. Richard Dawkins (1986, ch. 2) illustrates this point vividly in his discussion of the many delicate engineering problems that must be solved for bats to navigate by echolocation.

The thrust of this section can be formulated as a simple argument that shows how genuine innovation is possible, how it is possible to break the idea that design is conserved, the idea that you cannot get more from less.

1 Evolution (under reasonably favorable conditions) is inevitable.

2 Those conditions are widely available.

3 Evolution is innovative.

4 Thus innovation is inevitable.

So a process or phenomenon that seemed impossible turns out to be inevitable, given that "the reasonably favorable conditions" are actually quite abundant, not at all rare.

The question then becomes: Why can we not "reverse engineer" biological evolution to determine its secret and "bottle" this discovery so as to use it to solve our own problems in a manner that is deliberate and directed, by contrast with biological evolution? Indeed, does not evolutionary theory itself take a large step in this direction with the BV+SR model?

Interestingly, Hegel and Marx made a roughly similar point about human historical development. Historical development, too, has been a remarkably creative process; and the most creative developments of all occur behind our backs, unrecognized by us. Hegel and Marx had in mind such developments as

[23]Supposing that this schema captures the core of evolutionary theory commits one to the so-called adaptationist paradigm, according to which all genuine design in biological nature is adaptive design, the product of adaptation. This view has been challenged from various quarters, such as the neutral evolution of Motoo Kimura (1983) and the critique of Stephen Jay Gould and Richard Lewontin (1979). It is also challenged wholesale by those who see design, including adaptive design, as emerging out of complex processes. My own position is flexible. I am prepared to acknowledge other factors, but I believe that adaptation is by far the most important, that without adaptation the other processes would produce little.

the emergence out of feudalism of modern capitalism, the nation state, parliamentary democracy, individualism, and bourgeois life.

Now such human and natural historical developments do not, of course, entail the existence of an underlying method or logic or Cunning of Reason (*pace* Hegel and Marx); and yet they should give us pause; for these developments are highly creative despite departing from what we may call the traditional, *human design model*. As Peirce (1877, §1) already observed, Darwin in effect introduced statistical-populational methods into biology, to which we add the feedback mechanism that makes sense of biological function (Wright, 1973). Darwin's solution to "the mystery of mysteries" is a process that is blind, parallel, and distributed. And history, for Hegel, is the product of a vast, parallel process of everyday human actions distributed over rather ignorant processors acting and reacting locally—that is, people like us! Hence my inclusion among the mottos of the fictional exchange between Schelling and Hegel from Robert Solomon's book of mock-interviews, *The German Idealists* (Solomon, 1981, p. 56).

> *Schelling*: But surely you do not want to give so much emphasis to the random contingencies of history!
>
> *Hegel*: Even random contingencies may yet have some sort of logic. ...

Few philosophers of science today would take seriously the details of Hegel's logic, and yet there may be a point here worth thinking about. Perhaps we should not restrict method to step-by-step, linear processes that move logically from point to point, in the old sense of logic. Perhaps we need a wider conception that is inconsistency-tolerant or that even makes creative use of inconsistency (as Hegel's logic suggests); whereas, in standard logic inconsistency is totally destructive.[24]

[24]Oddly enough, Popper himself makes positive use of inconsistency. The hypothetico-deductive (H-D) method has always done that to a degree, but it was Popper who emphasized that "error" (contradictions between our latest hypothesis and its test results) is good—a necessary feature of scientific progress.

Diderik Batens, Joke Meheus, Erik Weber, and their colleagues in the subfaculty of logic at Ghent are actively exploring nonclassical logics that are inconsistency-tolerant, ampliative, and dynamical in other respects. These are still recognizable as logics in the sense that they each have a semantics and a proof theory. Below I shall introduce an approach that goes beyond logic even in this liberalized sense. Although our approaches are very different, I don't think they are incompatible. I agree with the Ghent logic group that active human reasoning, even when rational, rarely follows classical logic. Thus an adequate account of reasoning cannot avoid employing nonclassical patterns. Second, I agree with Campbell that BV+SR processes underlie all cognition involving novelty, so BV+SR cannot be avoided either. Therefore, the two approaches must be compatible. An adequate account must satisfy both constraints. My present view is that BV+SR, much of it at the subconscious level, underlies the patterns that the new Ghent logics display. We expect higher-order patterns to emerge, some of which are accessible enough to achieve general normative value. Classical logic does that, too, but at such a crude level that it fails to reflect ordinary reasoning. The Ghent team is able to capture far more of the rich detail of actual thinking. (Many of their papers are available from their web site: http://logica.ugent.be/centrum/writings/.) Actually, their own research smacks of BV+SR, which is not surprising. They explore a variety of logics of various kinds and select those that are interesting for various reasons.

5. Two Objections

Having raised our hopes by re-examining the evolutionary refutation of general method, it is only fair that we consider more seriously the Popper-Campbell objections. One way of putting their position is that since science develops, and can only develop, by means of an evolutionary process exemplified by Darwin's, that there cannot possibly be a *method* of innovation. For our best example of such a process—biological evolution—is purposeless and blind. It is a product of zillions of small chance events transformed by a somewhat cumulative statistical process into the emergence of new life forms. Thus the evolutionary character of science is the *reductio ad absurdum* of the idea of a method of scientific discovery. In fact, biological evolution nicely illustrates the disjunction of discovery and justification, for there is something in evolution that corresponds to criteria of justification, namely relative fitness tests. Variants blindly produced by a trial-and-error process are kept or rejected according to whether or not the result meets the fitness criteria.[25] Clearly, it does not matter how they are produced, only whether and why they are retained. In short, the evolutionary model calls to mind discovery by monkeys sitting at typewriters.

Second, and contrary to the impression given by the Hegel passage above, human problem solving, human design, human innovation, is not at all like this. Human efforts are consciously directed toward problems or ends specified in advance and hence are heuristically motivated. People can employ abstract representations and manipulate them logically and mathematically. People can try out ideas in the abstract and protect them during their developmental stages. People can combine ideas from very different species of things. And so on. The human design model is a quite different model of problem solving.

Just as the fact of biological evolution constitutes an existence proof (biological evolution, the emergence of more design from less is possible because it is actual), so, defenders of the human design model will say that the fact that human history has produced many examples of innovative design is also an existence proof that demonstrates the applicability and robustness of the human design model. But let us look more closely at the human design model.

Must not defenders of classical logic as the only account of reasoning either deny that ordinary human thinking, say in trying to figure out something, is *reasoning*—a distinction parallel to that the logical empiricists and Popper (1934) made between context of discovery and context, or logic, of justification; or else claim that ordinary thinking and problem solving is deductive, or deductive plus standard inductive? In my opinion, the latter alternative makes the same mistake as those who confuse original discovery processes with what I call discoverability or generatability (see Notes 10 and 14). It is to confuse the logic of the "final" product with the reasoning of the process that produces that product. It was strange to see fallibilist philosophers of science give almost exclusive attention to the logical structure of the final products, given that these products typically do not last long but are merely temporary stabilities in a larger, ongoing process.

[25]This biological process is very far from totally blind or "random", of course. It is highly constrained variation. Rabbits breed other rabbits, not giraffes or elm trees or washing machines. In fact, they breed rabbits of the very same species, and offspring that closely resemble their parents.

Once this model in place, we appreciate that *the God design model* is just the human model blown up to infinite size. Paley's point was that human bodies are machines, artifacts, just as the watch is, but so vastly more complex that they could only have been made by a Great Artificer. Whether or not Paley was justified in extending the watch example to a proof of God's existence, his heath-walker was surely justified in concluding that the watch-like object in the path was indeed a humanly made watch, constructed intentionally according to a detailed plan, rather than a chance conjunction of the elements.

Still emphasizing the differences, we note that Darwinian evolution amounts to the discovery (anticipated by Hegel and others) of a control theory quite different from either of the classical control systems mentioned at the end of §3, the linear causal-chain model and the logical model. Although Darwin's account is causal, it is not linear.[26] It involves a subtle, indirect sort of feedback, many decades before Norbert Wiener (1948) and others clearly articulated the idea of feedback control. Nor need it be implemented as an information-theoretic or logical system, although it is sometimes useful to represent it as such. After all, biological nature is hardly a symbol system. Regarded as an information-processing system, an iterated BV+SR system is very "noisy"—because of the large role of chance contingencies. In standard logical and information-processing systems, noise is something to be ignored or eliminated as far as possible. But if Campbell and company are correct, fully to eliminate such noise would throw out the baby with the bathwater when it comes to innovative potential. For it is precisely the BV part of the BV+SR process that permits creative breakaways from the current constitution of the system.

Gerald Edelman, who applies selection theory to the ontogenetic development of the central nervous system ("neuronal group selection"), emphasizes this point. The problem with information theories, he says, is that they are too precise to be creative (Edelman, 1987, chs. 2 and 3). Max Delbrück raised the point to a "principle of measured sloppiness"

> If you are too sloppy, then you never get reproducible results, and then you can never draw any conclusions; but if you are just a little sloppy, then when you see something startling you [nail] it down.[27]

A more common way to distinguish learning theories, including theories of innovation, is to classify them as either *providential*, *instructionist*, or *selectionist*. A providential account explains learning, achieved fitness, or design in a direct, creationist sort of way. The knowledge or other design is simply given providentially by a divine being. We are born with innate epistemic gifts such as innate knowledge that call for no additional explanation. This is one version of

[26]For a popular account of Darwin's innovation from a control theory perspective, see Cziko, 2000.
[27]Quoted by Rheinberger (1997, p. 78).

an epistemic "given". Instructionist theories posit epistemic givens of a different kind: e.g., simple empiricism, which enables us to read the truth directly off nature, and the somewhat more complex theory of passive induction, whereby mere repetition in nature establishes habits of mind and hence commitments to general claims and practices. Simple, instructionist empiricism models the mind as a wax tablet upon which the forms impress themselves. Historical examples are Aristotle's theory of in-form-ation and Hume's psychological theory of habit formation and induction, which Popper famously criticized.[28] This is one form of what Wilfrid Sellars (1956) dubbed the "myth of the given".

Darwin and Wallace discovered the third, selectionist (BV+SR) theory of learning, and many now espouse it in some form as a solution to the problem of how innovative inquiry is possible at all.[29] This model posits a mechanism of variation, a mechanism of selection, and a mechanism of retention or transmission to the next generation, operating in a reasonably stable environment with selection pressures. The key idea is that an evolving population can and regularly does produce variants *fitter than any that previously existed* (Altenberg, 1994). Novel design evolves—*emerges*—from a multi-stage process of cumulative adaptation. Novel design is created from nothing, not literally *ex nihilo* but in the sense that *more design emerges from less, contrary to the creationist model.* This means rejecting such conservation principles as that you cannot get more knowledge from less or more design from less.

6. The Triumph of the Darwinian Method?

What can we reply to the objections of the previous section, as the next stage in our ongoing dialectic? Let us take them in reverse order.[30]

The human design model is providential, instructionist, or both. Insofar as the human designer fully and consciously controls each step of the process toward a pre-specified goal, the model is inherently limited to what the human designer can presently imagine. The model cannot explain how it is possible for a work of art or a research paper to contain more than the designer deliberately

[28] See, e.g., Popper, 1963, pp. 42ff.

[29] See the works by Popper, Campbell, Dawkins, Edelman, Hull, Plotkin, Dennett, Simonton, Kantorovich, and Cziko, cited in References. When extended to all learning and adaptation, the thesis is called universal Darwinism, or universal selection theory. To establish universal Darwinism, we need to show that it is both sufficient and necessary for novel design or adaptation. The aforementioned existence proofs establish sufficiency, and the refutation of providential and instructionist theories (more fully discussed in Nickles, 2003a) establishes necessity, or at least that no other known theory can account for novel design. Neither necessity nor sufficiency, as I am using the terms, denies that factors other than BV+SR are involved in creating novel design, e.g., physical constraints. Accordingly, the sufficiency claim must be qualified. From the beginning, Darwin and all BV+SR theorists have acknowledged that certain broad environmental and physico-chemical background conditions must be in place for evolution to occur.

[30] The title of the present section alludes to Ghiselin (1969).

put into it.[31] Almost by definition, *fully* intelligent planning cannot be creative, since the end product and the process for producing it are fully anticipated and hence routine. It is only a matter of implementing a design that one already has. (As if it is the construction workers rather than the architects who are truly creative.) Thus the expressions 'fully intelligent novel design' and 'fully intelligent creativity' are at least as oxymoronic as 'chance-based novel design'. By contrast, the BV+SR paradigm allows for lateral shifts in the inquiry process, whereby we don't get what we thought we originally wanted but instead want what we get. (Or future interpreters examining our work from within their own inquiry agendas will want what they get out of it).

Notice, then, that the God design model does not really explain where innovative design comes from, for the Judeo-Christian God, as an infinite being, already contains all possible design. Let us say "infinite design", for convenience. As Dennett (1995) points out, the religious tradition attempts to explain less design in terms of more design. Postulating God, as the argument from design does, postulates infinite design as present from the beginning, as if all that design is somehow self-explanatory, the unexplained explainer. Thus, ironically, in this crucial sense, God's activity is not creation *ex nihilo*, whereas Darwinian evolution's is! There is also something methodologically discomforting about the argument from design. Postulating infinite design to explain a relatively little design is to use a very big cannon to kill a tiny fly.

As hinted above, the old view that you cannot get more design from less, that there must be at least as much design in the designing process as in the designed object, amounts to a conservation principle, call it the Principle of Conservation of Design. Descartes explicitly committed himself to a similar principle, but he was just following tradition at this point. What the evolutionary model shows is that this principle is false. We can get more design from less.

Third, the claim that the creativity of human history is an existence proof that validates the human design model founders on an ambiguity between evolution as a fact and evolution as a theory or specific mechanism. In §5 we specified the mechanism of evolution (very schematically, but a great deal of detail is now known and could be filled in). We have no similar specification of the mechanism behind the human design model when it is applied to novel problems as opposed to tasks that are consciously preplanned in every detail.

Many "mechanisms" have been suggested, of course, typically exceedingly vague ones that have the failings of providential and instructionist theories. For example, there are romantic models that postulate the existence of geniuses

[31] Here I mean something other than the familiar point that no one can see all the deductive consequences of one's position. There is a related issue in literary theory, namely whether the "meaning" of a text is identical with the author's intentions. My answer is "no", for literary critics regularly illuminate rich texts in ways that the author did not intend. See Bloom, 1973 on strong reading.

with special ability to see into the structure of the universe and religious models that involve inspiration from God. But, again, if Einstein got his ideas from God, then we should not say that Einstein was creative. And to postulate the existence of individuals with a special faculty of precognition or clairvoyance begs the whole question of how we get more knowledge from less, more design from less, as Campbell himself has insisted. Many analysts today reject all non-natural faculties traditionally attributed to human beings, including the rationalist faculty of reason with its capacity of intellectual intuition, the empiricist faculty of veridical, direct perception of the world, and any source of Kantian synthetic a priori knowledge.

Furthermore, as Campbell himself emphasized, the closer we look at actual cases of novel human design, the less it appears that the human design model is applicable. Much recent work in science studies supports this claim for the domain of the sciences. The closer we look, the more human innovation resembles an evolutionary process. This is apparent in everything we do in science, technology, the arts, and everyday life. It is no accident that the earliest automobiles resembled horse carriages and that the earliest TVs resembled big radios. It is no accident that a research paper undergoes many drafts and that the ideas in it, if they attract substantial positive notice, subsequently undergo several layers of refinements. Moreover, the most creative human inventions that we find in history were largely blind. No individual or group of people could have sat down in the year 1300, say, and said, "Let's create new forms of painting, namely, fully representational works in linear perspective. And while we are at it, let's go on to design a new economic system with a market economy and a corresponding new political order centered on nation-states having the form of parliamentary democracies." In fact, it was not until 1776, hundreds of years after the emergence of early capitalism in the city-states of northern Italy, that Adam Smith was able to articulate, in fairly clear terms, the mechanism of capitalism. Even at the individual level, search for a solution to a new problem is partially blind.

But what about deductive logical arguments? Surely the human design model gets a foothold there, a basis for expanding into many domains? Well, consider finding a proof to a theorem of geometry or logic.

> [F]rom a logical standpoint the processes involved in problem solving are inductive, not deductive. To be sure, the proof of a theorem in a formal mathematical or logical system [such as Logic Theorist] is a deductive object; that is to say, the theorem stands in a deductive relation to its premises. But the problem-solving task is to discover this deduction, this proof; and the discovery process, which is the problem-solving process, is wholly inductive in nature. It is a search through a large space of logic expressions for the goal expression—the theorem. (Simon and Lea, 1990, p. 26)

Any program that requires genuine search, any strategy, such as generate-and-test, that includes test operations, contains an element of BV+SR (although not necessarily evolutionary BV+SR in the full-blooded sense).[32]

We cannot take even deductive reasoning as an innate, God-given human ability that needs no further explanation. At the microcognitive level, various naturalistic theorists treat recognizing deductive structures as a matter of pattern matching, that is, trial-and-error fitting, much of which occurs at the subconscious level (Johnson-Laird and Byrne, 1991, Margolis, 1987).

If the preceding line of thought is correct, then we have another great irony. The human design model is not an alternative to the evolutionary BV+SR model, and superior to it. On the contrary, human design, whenever it is applied to innovative work, is underlain by the BV+SR model. The traditional view is upside down. At bottom, the human design model, insofar as it can be genuinely creative, must rely on BV+SR processes. If the human design proponent refuses to accept this, then we must say that the human design model (bereft of any chance processes) is simply wrong, even question-begging, logically fallacious. I leave it to the reader to ponder the implications for the God design model.

Campbell (1974a) develops this point at some length. Insofar as human problem solving is directed rather than random, it is on the basis of knowledge (or beliefs) previously achieved, and these achievements were in their turn the product of prior BV+SR processes. And insofar as the heuristic guidance runs out, we can only proceed blindly.[33] A gopher scanning its surroundings for predators obviously has a highly developed capacity (acute vision, hearing, smell) to detect threats at a distance. This capacity is itself the product of prior evolution, not a non-natural "given". Yet even with this capacity, the gopher must still scan, and it is a matter of chance whether or not there is a predator to spot at, say, the two o'clock position at this particular moment. In that sense, even its sighted scanning is "blind" search. After all, if it already knew where and when the predators were, no search, no visual scanning, would be necessary.

Campbell, Dawkins, Dennett, Edelman, Hull, Plotkin, Cziko, and other universal evolutionists contend that evolutionary processes are the *only* processes we know capable of producing genuine innovation, more adaptive design from less. Evolution is the only game in town. To establish this requires arguing

[32]We can thus extend an important argument of Simon. (1) All inquiry, including that involved in scientific discovery, is problem solving. (2) All problem solving involves search. (3) Therefore, all inquiry involves search. If we add as another premise (4) All search requires BV+SR, then we may conclude: (5) All inquiry requires BV+SR. Actually, I think premise (1) is a little strong. Insofar as we want what we get, we are in the position of having a "solution" in search of a precise formulation (or reformulation) of a problem. But this still requires search, after all.

[33]Critics commonly object at this point that human inquiry is not blind in the sense that biological evolution is. Quite true (as I noted in §5), but the objection misses the point. The claim is not that human inquiry is analogous to biological evolution, which is one specific kind of BV+SR process. But it is nonetheless blind in the sense that Campbell specifies.

for both the existence and the uniqueness of evolutionary processes as processes capable of producing genuine innovation. The preceding paragraphs have sketched both parts of this argument. The existence proof of evolution constitutes one part, and the argument that any alternative to a cumulative, trial-and-error process is question-begging provides the other.

Having dismissed the second objection of §5, we are back to the first, Campbell and Popper's claim that evolution is the last nail in the coffin of methodology of discovery or innovation. One cannot generate a method of discovery out of monkeys at typewriters!

I reply: Yes we can! A primary problem with the monkeys at typewriters is that there are not enough monkeys! (Just as, in Paley's day, Darwin's theory would have been dismissed as a joke, for the obvious lack of time if for no other reason.) And the products of their work need to be selected and transmitted cumulatively, as in evolution. In this case, "scaling up" can make a huge difference. And scaling up is exactly what the new AI development called evolutionary computation does. Here is precisely the attempt to reverse engineer evolution, to "bottle" it, that is, to articulate it as a problem-solving method, and to apply it to the problems that interest us.

7. BV+SR: Madness or Method?

BV+SR has been touted by some of its primary philosophical supporters and critics alike as the very antithesis of method. As far as I know, it was first ridiculed by Jonathan Swift's satire, *Gulliver's Travels*, as "the Laputan method of making books". What could be less methodical and less efficient than blind trial and error? If science operates by BV+SR, then, it is claimed, there can be no scientific method, or at best a minimal logic of justification that keys on the mechanism of selection.[34] With such a process, above all, a sharp distinction between context of discovery and context of justification would seem to be in order, for the context of discovery will contain an unusually large number of false starts and contingencies irrelevant to final testing and justification.[35]

Let us consider some problem-solving search procedures.

1 The so-called British museum algorithm covers everything. It searches in a "brute force" manner through the entire search space or problem-solution space. The trouble with this strategy is that the boundaries and

[34]Philosophers themselves have taken little interest even in the realistic implementation of selective-retention mechanisms in actual communities. For example, Popper would have us retain all conjectures that are not yet refuted, whether or not they are highly corroborated or heuristically productive. For him, officially at least, only explicit refutation removes a hypothesis from the "table".

[35]Yet the thousand small episodes of testing or evaluation that mark them as false starts also belong to "logic of justification". I do not deny the existence of various distinctions between context of discovery and context of justification, but I do claim that they have often been misused.

internal structure of the search space are largely unknown insofar as one is working at the frontier of research, so a systematic search is impossible. Moreover, even when such domain knowledge is available, the method frequently becomes hopelessly uneconomical because of the sheer size of the domain. Heuristic shortcuts are often faster.

2 A second[36] sort of search lies at the opposite extreme: a single, blind or "random" effort to find an item of the desired description in a large search space—a needle in a haystack. Here the problem-solving efficiency is very low, because the probability of success diminishes rapidly to zero as the search space becomes larger. This is the point of the creationists' (mistaken) criticism that Darwinian evolution is tantamount to a single explosion in a print shop producing the works of Shakespeare. It also corresponds to taking down a single volume from Jorge Luis Borges's Library of Babel in searching for, say, Dostoevsky's *Crime and Punishment.*[37]

3 A third sort of search involves multiple tries at the answer, in series or in parallel, as represented by monkeys at typewriters, whose work is checked (and either retained as correct or discarded) when a typescript of the desired length is completed. (Some writers term this rather than brute force the British museum model, since the monkeys are imagined to reside in the basement of the British Museum, perhaps slowly replicating all of its works, one by one.) This model mitigates the explosion-in-a-print-shop objection by allowing multiple explosions. If there are enough monkeys (or one monkey trying again and again over a very long time), then the probability of producing a workable problem solution increases accordingly, but so does the investment in the effort. This strategy corresponds to several Borgesian librarians pulling down one volume each, or to a single librarian pulling down multiple volumes, entirely at random. It perhaps also corresponds to a Popperian scientist formulating a series of hypotheses.

4 A more economical strategy is to allow for a big dose of serendipity. Instead of requiring the solution to one big problem, stated in advance, check instead for any coherent or sufficiently interesting manuscript, whether previously anticipated or not. Prespecifying a goal and rigidly sticking to that research plan not only reduces the probability of hitting something interesting but also limits the innovation to what we currently

[36]These paragraphs sketch a series of possible search strategies. My discussion is far from exhaustive.
[37]See Borges, 1954. For discussion see Dennett, 1995, p. 107 and Kelly, 1994, pp. 258ff.

think we know, or can plausibly imagine. Such a constraint immediately dulls inquiry. Early Newtonians could not imagine electromagnetic fields, nor could early geneticists imagine the molecular genetics of a few decades later.

5 Far more economical still is to permit cumulative trial and error, cumulative adaptation, as in biological evolution. Here the BV+SR proceeds in incremental steps over a long period of time (tiny sparks, if you like, rather than big explosions), adaptively reacting to the local conditions that exist at that time rather than working directly toward some distant goal. This is the point that the Creationist explosion objection misses, as does the extreme romantic insight model of discovery. Again, it is wrong to see efficient inquiry as aiming to achieve a *major* new problem solution in one big jump to a predesignated goal. Rather, most progress consists of adapting solutions already available to more local problems. And, again, it is a matter of wanting what you get rather than of getting what you (think you) want.[38]

This last strategy, although seemingly far from methodical, is far more efficient, adaptive, and intelligent for a great many domains.[39] Thus methodologists should find it interesting.

Equally interesting is that the BV+SR process itself can be characterized algorithmically, in a suitably broad sense. It is useful to distinguish what we may call traditional, "vertical", goal-directed algorithms, which are guaranteed to produce a prespecified result by means of a series of rigorously indicated steps, from "lateral" algorithms, which also consist of a series of well-defined steps but which do not guarantee arrival at a predesignated goal. Since BV+SR processes can be defined in this manner, they are algorithmic in this broad sense.[40]

---

[38]It is odd that many experts agree that science is not wending its way to a preordained goal, at least not one clearly knowable in advance by us (else we would already be there); yet many of these same people still employ the human design model as the obvious one to use. This statement is true of several traditional AI experts as well as philosophers.

[39]A leading example of "inquiry" driven by BV+SR that efficiently solves the Meno problem of recognizing and responding to novelty is the vertebrate immune system. This system evolved phylogenetically, of course, but it also employs BV+SR processes that respond to new antigens within a single lifetime, indeed, within a few minutes. This ontogenetic BV+SR process shows how a small, finite system of inquiry can have such an enormous response domain. See Cziko, 1995, chap. 4 and Schaffner, 1980, pp. 175ff. Schaffner, however, employs Jerne's and Burnet's development of theories of the immune system in the service of the human design model of inquiry rather than analyzing the latter itself in BV+SR terms.

[40]See Dennett, 1995, pp. 48ff, also Koza and others cited below. The concept of algorithmic procedure is currently evolving toward greater breadth of application, in somewhat the manner of the concept of mathematical function in the 18th and 19th centuries. I shall not defend this claim in detail here but will provide an example below.

Thus, surprisingly, BV+SR is beginning to resemble method rather than madness.[41] If a process can be produced by an algorithm, then it can be modeled by AI. And, in fact, AI provides another existence proof, this time that powerful, efficient AI methods based on luck are possible. They are possible because they are actual. An entire field of research, evolutionary computation, already exists. Here I shall discuss *genetic algorithms* (GAs),[42] whose performance can be demonstrated even on a personal computer. This existence proof is the AI counterpart of the biological existence proof that shows that BV+SR processes can be incredibly creative problem solvers.

Even those of us who are not "PC positivists" should appreciate that, over the past half century, computer science—AI in particular—has been a primary locus of explicit methodological thinking. But unlike philosophers and historians, computer experts *must* take totally seriously issues of computability, computational economy, and pragmatic workability. Traditional artificial intelligencers, working within the human design model of problem solving, were initially as skeptical as the philosopher "friends of discovery" that BV+SR could be embraced as a powerful method rather than as anti-method. To those of us working on logics and heuristics of discovery, constrained search, etc., it initially seemed crazy to suggest that trial and error could be methodized. As I noted above, 'methodological luck' sounds oxymoronic.

Moreover, the basic idea behind GAs also sounds crazy at first. For how could randomly recombining lines of computer code or segments of computer programs possibly produce anything but uninformative failure? Yet work on GAs, still undergoing rapid development, has already produced remarkable results. It was John Holland who first clearly established the existence of productive genetic algorithms in the 1960s (cf. Holland, 1975, 1995), although Alan Turing already anticipated the approach. Holland used fixed-length, chromosome-like strings of binary code and bred them. Melanie Mitchell and Stephanie Forrest (1995, p. 268), two former students of Holland, outline the process.

> A simple form of the GA ... works as follows.
>
> 1 Start with a randomly generated population of chromosomes (e.g., candidate solutions to a problem).
>
> 2 Calculate the fitness of each chromosome in the population.
>
> 3 Apply selection and genetic operators (crossover and mutation) to the population to create a new population.

---

[41] Thus the title of the book in which Nickles, 1992a appears—*Science: Between Algorithm and Creativity*—expresses a false dilemma. It is hard to imagine a process more creative than biological evolution, a BV+SR process; yet such a process is algorithmic in a broad sense. Broadening the concept of algorithm enables us to have our cake and eat it too!

[42] At the time of the conference (1998), evolutionary computation was not nearly as widely known to philosophers as it is now. Thanks to Paul Teller for putting me on to genetic algorithms many years ago. A good introductory text is Mitchell, 1996. A standard text is Goldberg, 1989.

4 Go to step 2.

> This process is iterated over many time steps, each of which is called a "generation." After several generations, the result is often one or more highly fit chromosomes in the population.

The "chromosomes" are coded to represent potential problem solutions.

Instead of chromosome-like symbol strings, John Koza, another of Holland's former students, breeds populations of computer programs. Accordingly, he terms his approach *genetic programming*. His results may be found in four large, recent volumes (Koza, 1992, 1994; Koza et al., 1999, 2003), where he studies the application of GAs to dozens of interesting problems. His is just one of many different approaches to evolutionary computation; but his work serves as a good example, and he is quotable.

Writes Koza in the first volume of the series:

> I describe a single, unified, domain-independent approach to the problem of program induction—namely, genetic programming. (Koza, 1992, p. 3)

In the second volume, he describes the aims achieved in the first:

> *Genetic Programming* ... proposed a possible answer to the following question, attributed to Arthur Samuel in the 1950s:
>
> > How can computers learn to solve problems without being explicitly programmed? In other words, how can computers be made to do what is needed to be done, without being told exactly how to do it?
>
> *Genetic Programming* demonstrated a surprising and counterintuitive answer to this question: computers can be programmed by means of natural selection. In particular, *Genetic Programming* demonstrated, by example and argument, that the domain-independent genetic programming paradigm is capable of evolving computer programs that solve, or approximately solve, a variety of problems from a variety of fields.
>
> To accomplish this, genetic programming starts with a primordial ooze of randomly generated computer programs composed of the available programmatic ingredients, and breeds the population using the Darwinian principle of survival of the fittest and an analog of the naturally occurring genetic operation of crossover (sexual recombination). Genetic programming combines a robust and efficient problem-solving procedure with powerful and expressive symbolic representations. (Koza, 1994, p. 1)
>
> ...
>
> In practice, the genetic algorithm is surprisingly rapid in effectively searching complex, highly nonlinear, multidimensional search spaces. This is all the more surprising because the genetic algorithm does not have any knowledge about the problem domain except for the information indirectly provided by the fitness measure. (Koza, 1994, p. 27)

Koza's method is to start from a randomly generated population of computer programs constructed from an initial set of functions, certainly including the basic arithmetical functions and a conditional branching operator, from which many other functions can be constructed by the system. In early versions, the programmer specified the input variables or *terminal sets* and the *function sets*—the set of basic functions that might be useful. Later versions contain pre-specified function and terminal sets, so the programmer need not even do this. Each generation is tested for fitness and then probabilistically "bred" to produce the next generation. Fitter individuals have a higher probability of being bred. Since computer programs are hierarchical, tree structures of different sizes, two programs can "mate" productively. The breeding operation consists of crossing pairs of programs at randomly chosen nodes so that they exchange branches at that point while remaining legitimate programs.[43] The "sexual" mixing maintains the diversity of the population. Koza's team often finds a good problem solution within thirty to fifty generations, over a population of 500 to 1000 individuals. However, sometimes larger populations, longer runs, and multiple runs are necessary. Given the nature of the process, the GA does not solve the problem in exactly the same way, or in the same number of generations, on each "run".

The problems solved are not "toy" problems such as tic-tac-toe or the tower of Hanoi but practical engineering and scientific problems. Examples from the first volume include the problem of balancing a pole on a moving platform, backing a truck into a confined loading dock, animal foraging economies, optimal control problems, planning, curve fitting and function identification (symbolic regression), and image compression. The systems described in Koza's later volumes solve more complex problems by, in effect, defining their own new functions and proceeding in a hierarchical manner (calling and thereby reusing subroutines), and by giving programs the flexibility to change their architecture on the fly. For example, automatically defined functions (ADFs) enable the system more efficiently to exploit regularities, patterns, and symmetries that it notices in the problem domain, and hence to tackle "regularity-rich problems". In effect, the program decomposes such problems into more manageable subproblems and then reassembles the results; or, if you like, it implicitly alters the problem representation (Koza et al., 1999, p. 68).

Koza employs additional analogues to biology, notably gene duplication and deletion. These techniques are now being applied to analyzing large databases of the kind involved in molecular biology, e.g., in connection with the protein folding problem. Volume III features the design of analog electrical circuits and represents the first success in automating analog (as opposed to digital) circuit design, an intensive task previously accomplished only by experienced, insightful human engineers. Koza reports many automated discovery results

[43]Some approaches introduce an analogue of mutation in randomly chosen individuals at this stage, usually point mutations.

that are as good or better than the published human results (Koza, 1994, p. 16 and chap. 18; Koza et al., 2000), including several patented electronic filters. In fact, Koza's group has now filed for several patents of their own. Volume IV tackles still more difficult problems of synthesizing controllers, circuits of various kinds, metabolic pathways, and numerous other things. In fact, Koza's approach is spawning an entire industry.

The work described in the four volumes progresses toward Samuel's goal by permitting the computer user to specify problems to be solved at ever-higher levels of description. It thus reinforces the claim that you can get far more out of a computer than you put in. Koza calls the present version of his program "the Genetic Programming Problem Solver (GPPS)", a name reminiscent of Newell and Simon's GPS.

> GPPS is intended to provide a general-purpose method for automatically creating computer programs that solve, or approximately solve, problems. GPPS uses a standardized set of functions and terminals and thereby eliminates the need for the user to pre-specify a function set and terminal set for the problem. In addition, GPPS uses the architecture-altering operations to create, duplicate, and delete subroutines and loops (and, in GPPS 2.0, recursions and internal storage) during the run of genetic programming. Since the architecture of the evolving program is automatically determined during the run, GPPS eliminates the need for the user to specify in advance whether to employ subroutines, loops, recursions, and internal storage in solving a given problem. It similarly eliminates the need for the user to specify the number of arguments possessed by each subroutine. (Koza et al., 1999, p. 14)

In Newell and Simon's GPS, the overall strategy was to move point-to-point in solution space, in serial fashion, by means of logical and heuristic rules, and to do so in a way that mimics human protocols for the same problem. By comparison, Koza's system is a non-logical (in the Newell-Simon sense), highly parallel process of Darwinian competition among mutually incompatible alternatives.[44] Moreover, problems are not represented explicitly. This is a step toward handling the "ill-structured", poorly defined problems typically found at the frontier of science and other creative enterprises.[45]

How then are problems given to GPPS? They are specified implicitly, by means of the preparatory steps necessary to set up the GA. Most relevant here are the coding scheme (necessary to map the individual programs in the population to problem solution candidates such as an electric circuit with specifically valued

[44]The preceding sentence should not be taken to deny that serial processing is often sufficient and even necessary to solve a problem. Also, Koza et al. (1999, p. 380) point out that their approach sacrifices computational efficiency to generality; but so, after all, does biological evolution.

[45]For ill-structured problems, see Simon, 1973. See also my constraint-inclusion model of problems (Nickles, 1981), which considers some historical cases of ill-defined problems. Of course, 'ill defined' in this sense is still different from 'implicitly defined' in the evolutionary biological sense or even in Koza's sense unless the known constraints are explicitly represented in the fitness evaluation function.

capacitors, resistors, and inductors in a particular topological configuration) and the fitness criterion. There is also a stop rule that informs the program when to stop and display its best individual solution. This may take the form of a success predicate that must be satisfied by some individual in the population. However, the specific strategy for solving the problem remains unspecified. "Evolutionary methods have the advantage of not being encumbered by preconceptions that limit the search to familiar paths" (Koza et al., 2000, p. 123). In effect, the problem is specified by furnishing the system with a goal, set out in terms of a set of constraints, some minimal resources for reaching the goal, and nothing else.

There are still major differences from biological evolution, of course, since nature imposes far more diversified fitness criteria that, in effect, implicitly specify several possible problems at once, with no explicit stopping rule. Actually, problems are not given externally to nature in *any* form, so to speak of nature as a problem solver is to employ an even more implicit attribution of a problem than Koza's. Genetic programming is more analogous to artificial selection than to natural selection in the wild.

Can we imagine developing systems that are closer to wild nature? A critic may complain that such an exercise would be pointless, because we surely must specify a goal if we are to use such a device for our own purposes. Moreover, we already have available for study zillions of truly unconstrained problems solvers, where we, ourselves, must reverse engineer the problem as well as the solution—namely, biological species! So developing analogous artificial systems would be a wasted effort.

But this objection is a mistake, for biological evolution employs only a limited set of BV+SR processes. In other words, biological BV+SR occupies only a small region of the total space of BV+SR processes. Darwin and Wallace were first to discover the creative power and persistence of BV+SR processes, but we must not identify all BV+SR processes with the biological ones or with direct analogues to them. We can imagine future artificial systems left to run on their own and to compete in various ways (as in the transition from AI to AL = artificial life), whence we do indeed face the potentially fruitful task of reverse engineering the problems as well as the solutions. In fact, to some degree investigators are already doing that. Already with present-generation evolutionary computation, we face the problem of reverse engineering the solutions—trying to understand their point.[46]

[46]The most common objection to extending evolutionary (BV+SR) processes to individual learning, social evolution, and such is that it is not exactly analogous to biological evolution. But as Campbell, Hull, Dennett, and others have repeatedly stressed, this objection completely misses the point. An evolutionary explanation need not be biologically reductivist, even by analogy. Much of the point of current BV+SR research in AI and elsewhere is to explore non-Darwinian types of BV+SR.

In fact, GPPS itself remains too close to biology in some ways. It is too directly competitive to incorporate all of the tricks that human problem solvers have available. For instance, human problem solvers can deliberately (not randomly) save from elimination a fatally defective model that nevertheless seems promising for future development or that is desirable for other reasons. In other words, we can look ahead in the way that machines so far cannot. Of course, GPPS.12 may be able to incorporate within its fitness measure more subtle features of heuristic appraisal that narrow this gap. And its parallelism already goes far to address this problem.[47] Combining this sort of approach with case-based reasoning may bring further progress.[48]

8. The Generality Question and the NFL Theorems

The traditional scientific method, including method of discovery, was supposed to be general, domain neutral, and powerful. But as we saw in §3, AI experts claim to have learned in the 1970s and 1980s that power can be purchased only at the expense of generality and domain-neutrality. This conclusion is also supported by the aforementioned "No Free Lunch" theorems, which imply that for a general induction rule to work better than the average rule in a particular world, it has to be tuned to that world, that is, to incorporate some empirical domain knowledge of that world or specific domain. Time will tell how much progress evolutionary computation experts make in this direction. Moreover, no precise consensus has yet been achieved as to what the NFL theorems mean. Nonetheless, the theorems appear to put serious limits on generality aspirations.

Must not the NFL theorems hold for genetic programming as well? Do Koza and his associates reject these claims?[49] They tout the domain-independence and problem-independence of the GPPS sort of "machine". Thus Koza seems to be returning to the old (pre-knowledge-based programming) idea that the relevant domain knowledge can be given together with a statement of the problem and need not be explicitly incorporated in the program itself. His programs completely lack the hundreds or thousands of content-specific production rules (if-then rules) found in standard, knowledge-based AI.

So how does genetic programming overcome the trade-off between generality and power? One key is to apply the previous point that genetic programming corresponds to Darwin's artificial selection rather than natural selection in that human beings set the goals and determine what the criteria of fitness are, in a problem-specific manner. Another is that we humans have to set up an appro-

[47] See the discussion of Holland below.

[48] Combining case-based reasoning with genetic programming is in fact the research program of Sushil Louis of the University of Nevada, Reno, for example. Consult the many papers available on his website (see References).

[49] I don't know that Koza's group has explicitly addressed the NFL theorems.

priate coding scheme so that the individuals in the competing populations can be interpreted by us as possible solutions to our problem.

> [T]he genetic algorithm does not have any knowledge about the problem domain except for the information indirectly provided by the fitness measure and the representation scheme. (Koza et al., 1999, p. 27)

> The preparatory steps of genetic programming are the user's way of communicating the high-level statement of the problem to the genetic programming system. The preparatory steps identify what the user must provide to the genetic programming system before launching a run of genetic programming. The preparatory steps serve to unmistakably distinguish between what the user must supply to the genetic programming system and what the system delivers. (Koza et al., 1999, p. 33)

> Finding a representation scheme that facilitates solution of a problem by the genetic algorithm often requires considerable insight into the problem as well as good judgment. (Koza et al., 1999, p. 20)

These remarks point the way toward resolving the generality puzzle. Obviously, Koza is not claiming that his systems solve problems by means of totally random search, which would be absurdly inefficient (Koza et al., 1999, p. 11). Their generation and selection components are highly constrained. Domain and problem specificity enter through the aforementioned preparatory steps. What Koza's research cleverly accomplishes is (1) a neat separation between the content-independent problem-solving system itself and the human input and (2) a remarkable reduction in the amount of human input necessary to assign the system a problem. How far this approach can take us remains to be seen, of course, but it is highly interesting from a methodological point of view.

My own take on the generality issue is this. I believe that Popper, Campbell, Dennett, and others are correct that a BV+SR process is needed to solve the problem of the growth of knowledge.[50] BV+SR programs are weak and inefficient compared to routine problem-solving procedures, but at the frontier of research stronger methods are simply not available to us. Thus we should expect many BV+SR methods to be more general than highly dedicated methods. Even at the frontier, however, we possess a modicum of domain knowledge. And that knowledge can inform our choice of a specific genetic program to use. At least we know enough to formulate an ill-structured or badly-posed problem. Now being able to formulate a problem in a highly specific manner already requires a lot of domain knowledge, so frontier problems will often be formulated more vaguely. Thus it is a progressive step that Koza's systems permit problems to be formulated in an ever-more high-level manner, with the missing specification to be automatically but conjecturally specified by the system.

Can a highly general, relatively domain-neutral method eventually result? Well, yes and no. "Yes", in the sense that a Koza-like system provides a

[50] And Plato's problem of the *Meno* (Nickles, 2003a, 2003b).

surprisingly general BV+SR "shell" or canonical methodological schema or strategy. But "no" in the sense that the shell must be specified to some degree before it becomes a useful problem-solving tool. Yet this is a surprisingly weak 'no', for the specification is given in the problem statement itself; it is not intrinsic to the system. And, again, the problem statement can occur at a high level that only implicitly defines the problem for the system. But on second thought, 'method schema' is redundant, since method is, by nature, at least somewhat general in application. After all, the H-D method is highly schematic. At any rate, particular BV+SR systems will be domain specific. Remember that Darwinian evolution itself does not consist in one, big, genetic algorithm. Rather, it consists of a large set of such processes, sometimes nested within one another, most of which are highly context-sensitive.

9. The Classical Discovery Program Revisited

With respect to our opening story, the BV+SR discussion reveals a new sort of argument from design. In the religious domain, the argument is negative, arguing against rather than for the existence of a transcendental creative entity or process. The same is true in the methodological sphere: there is no need to postulate a transcendental epistemological process, including a transcendental method. The design we find in knowledge enterprises, as elsewhere, has a naturalistic explanation. And, surprisingly, this explanation can be parlayed into an argument for the existence of a kind of *method* for producing novelty, only a method that we can access only partially at any time, a method still grander in scope than that imagined by Bacon and Descartes. All you have to do to create socio-cultural novelty is to gather together all of the cultural elements that exist at any one time, shake them up together repeatedly, and keep those new combinations that you want! For where else could novelty come from than by suitable recombinations and selections of available cultural resources? These cannot be limited to crude, mechanical recombinations, for metaphor, analogy, and other rhetorical tropes and interpretive devices are among the resources, and they can catalyze "chemical" combinations. So, again, what other source of novelty could there be?

Of course, we cannot explicitly muster all of these cultural resources at once, in one, big, centralized process, and then recombine them and select them in all possible ways. But the general method does exist. It is again just a matter of scale. These resources vastly exceed our ability to identify and control them in such a way. The Owl of Minerva flies out at dusk. So the novelty-producing process must proceed blindly, in the Hegelian manner (*sans* Reason), except for those limited areas where we can explicitly mount and control BV+SR processes, either formally or informally. For even a limited "shaking and baking" can produce novel combinations, and we may want some of what we get. Of course, such a process is going on constantly, subconsciously, in individual minds and also in communities and societies and their material resources, on many scales. Only rarely, by comparison, do we make it deliberate and explicit.

So what does it all come to? When we compare the advertised capabilities of genetic programming with the classical discovery program, we find a surprisingly good fit, given that we initially had no reason to expect any fit at all. Three or four centuries of devastating objections have totally discredited the classical program. It did attempt the impossible. But weaker programs of surprising generality are now becoming available to us. Here is a case where unwarranted optimism (read "hype") by the founders of modern methodology and, much more recently, by the computer science and AI communities has sustained an idea until it may, in an unexpected form, be starting to bear fruit. Evolutionary computing, now employed in thousands of technical articles each year, has suddenly become the most productive area of AI. Yet the BV+SR idea was supposed to be the very antithesis of method. A strange story indeed!

The BV+SR selectionist model of inquiry

- can provide a combined method of discovery (problem-solving) and justification.
- is algorithmic-computational in a broadened sense, and, in that sense, quasi-logical, a genuine method.
- is a routine capable of producing genuine novelty, a routine for producing nonroutine solutions, traditionally thought to be impossible.
- is quite general (Koza's large variety of problems), even universal (*if* we accept Campbell's thesis that *all* problem solving is underlain by BV+SR).
- partitions inquiry into a general, relatively neutral component and an "empirical" component.
- is nearly domain- or content-neutral, as neutral as we could expect given that a viable method
- must presuppose some substantive content or patterning of the universe.[51]
- is powerful: it solves difficult problems in an efficient, dynamical, adaptive manner.

[51]However, the method is not completely content-neutral, for no method will work in some worlds, e.g., the world of Borges's chaotic, unordered library. At no point does Koza suggest that genetic programming will work in all possible worlds or that it is always as good as some alternative method.

- is more powerful than the H-D method in being massively parallel and in not restricting itself to deductive testing.

- is more robust than the general method of hypothesis (which allows hypothetico-*inductive* as well as deductive inferences) in that a single run employs "the method of multiple working hypotheses" (Chamberlain, 1897) on a vast scale; and, once set up, it is easy to run the system several times. Roughly speaking, the traditional method of hypothesis is a BV+SR process writ small, in which only one hypothesis is tested at a time and in an all-or-nothing manner rather than by comparison with available competitors. While falling far short of the foundational justification of the classical discovery program, evolutionary computation thus better manages the problem of underdetermination of theory by fact.

- is as flexible as the H-D method in that it can easily be combined with other methods such as case-based and model-based reasoning.

- allows problems themselves to evolve and views problem solution as a matter of mutual fit between evolving problems and solutions.

Holland (1992, p. 66) makes this last point as follows:

> Pragmatic researchers see evolution's remarkable power as something to be emulated rather than envied. Natural selection eliminates one of the greatest hurdles in software design: specifying in advance all the features of a problem and the actions a program should take to deal with them. By harnessing the mechanisms of evolution, researchers may be able to "breed" programs that solve problems even when no person can fully understand their structure. Indeed, these so-called genetic algorithms have already demonstrated the ability to make breakthroughs in the design of such complex systems as jet engines.
>
> Genetic algorithms make it possible to explore a far greater range of potential solutions to a problem than do conventional programs. ...

Holland provides a nice image that illustrates the power of parallel processing (cf. Koza, 1994, pp; 27, 42). Imagine a biological fitness landscape in which each possible genome is represented as a column with a height proportional to the fitness of that particular genome relative to a given environment. Such a landscape is typically quite rugged, indeed, intricately so with multidimensional mountain peaks, valleys, ridges, gorges, and so on. Now a single investigator is like an ant with a hill-climbing heuristic exploring this vast space. The chance that it will find one of the high peaks is minuscule, since it almost certainly will get stuck on a local maximum or a ridge. This corresponds to the Baconian inductivist, the Cartesian deductivist, and also the Popperian hypothetico-deductivist, working alone.[52] By contrast, the selection theory invites us to imagine an entire population of ants, each sampling a region of the fitness space. Those ants that reach higher altitudes have a greater probability of breeding, with the result that successive generations focus the search increasingly on just those regions of the search space that are most promising.

We must remember that Koza's genetic programming is just one approach to the already large field of evolutionary computation. It is also worth noting that these approaches have other advantages that the founders did not attempt to incorporate in the classical discovery program. One counterintuitive feature is that they are methodical without specifying any particular path to the goal and in some cases without highly specifying the goal itself. Unlike traditional methods, which are almost synonymous with path-specification, BV+SR systems do not specify the One True Path to the One True Goal. For another thing, they are naturalistic, compatible with what we know today about the cognitive capabilities of humans and machines. Additionally, the processing is parallel and potentially massively parallel. These systems have competition built into them from the start. On the other hand, such a system cannot yet correspond to a genuine community of investigators.

These systems are still very far from achieving the capability of experienced human investigators, but they are approaching the stage at which they can be helpful assistants accomplishing specialized problem solving tasks. No one today wants to claim that we are close to possessing a general method of discovery or interesting innovation capable of postulating and justifying the existence of deep, novel theoretical structures. For that, the systems would have to be scaled up into new dimensions, perhaps by combining them with case-based and model-based reasoning employing "rhetorical" metrics for analogy and metaphor, a momentous task.[53]

Distant though the more optimistic vision for deep-theory generating machines may be, the existence of present generation computational systems already explodes some in-principle objections to machine discovery. Although complete realization of the classical discovery program is impossible, some lesser ambitions are not as impossible as they were once said to be.

If Bacon, Descartes, Newton, and Leibniz could have continued their "walk" through history into the future and could have visited one of our molecular biology or AI laboratories (for example), they would be amazed and delighted at the automated and semi-automated processing already in our possession and

[52]Since the H-D method does involve chance or luck on the side of hypothesis generation, we can regard it as a tediously slow BV+SR process in which the population is limited to one, two, or three hypotheses competing at a time.

[53]This is a momentous task even for human investigators. Kuhn dismissed Feyerabend's call for proliferation of deep theories on the ground that it often consumed all available resources to produce just one decent theory. Optimists can reply that the availability of machine assistance in the future can help to overcome this problem.

in daily use.[54] They would be especially surprised and puzzled at the creativity of BV+SR processes and even more surprised that "luck" can be methodized.

References

Altenberg, L. (1994). The evolution of evolvability in genetic programming. In Kinnear, K., editor, *Advances in Genetic Programming*, pages 47–74. MIT Press, Cambridge, Mass.

Bloom, H. (1973). *The Anxiety of Influence: A Theory of Poetry*. Oxford University Press, Oxford.

Bloor, D. (1976). *Knowledge and Social Imagery*. Routledge, London. Second edition: 1991.

Borges, J. L. (1954). The library of babel. In Borges, J. L., editor, *Labyrinths*, pages 51–58. New Directions, New York.

Brannigan, A. (1981). *The Social Basis of Scientific Discovery*. Cambridge University Press, Cambridge.

Campbell, D. T. (1974a). Evolutionary epistemology. In Schilpp, 1980, pages 413–463. Reprinted (along with other relevant papers) in Campbell's *Methodology and Epistemology for Social Science: Selected Papers*. Edited by E. S. Overman. University of Chicago Press, Chicago, pages 393-434.

Campbell, D. T. (1974b). Unjustified variation and selective retention in scientific discovery. In Ayala, F. and Dobzhansky, T., editors, *Studies in the Philosophy of Biology*, pages 139–161. Macmillan, London.

Campbell, D. T. (1997). From evolutionary epistemology via selection theory to a sociology of scientific validity. *Evolution and Cognition*, 3:5–38.

Chamberlain, T. C. (1897). Studies for students. *Journal of Geology*, 5:837–848. Reprinted under the title, "The Method of Multiple Working Hypotheses", in *Philosophy of Geohistory, 1785-1970*. Edited by C. Albritton. Dowden, Hutchinson and Ross, Stroudsburg, PA, 1975, pages 125–131.

Crevier, D. (1993). *AI: The Tumultuous History of the Search for Artificial Intelligence*. Basic Books, New York.

Cziko, G. (1995). *Without Miracles: Universal Selection Theory and the Second Darwinian Revolution*. MIT Press, Cambridge, Mass.

Cziko, G. (2000). *The Things We Do: Using the Lessons of Bernard and Darwin to Understand the What, How, and Why of Our Behavior*. MIT Press, Cambridge, MA.

Dawkins, R. (1986). *The Blind Watchmaker*. W. W. Norton, New York.

Dennett, D. (1995). *Darwin's Dangerous Idea*. Simon and Schuster, New York.

Edelman, G. (1987). *Neural Darwinism: The Theory of Neuronal Group Selection*. Basic Books, New York.

Ghiselin, M. (1969). *The Triumph of the Darwinian Method*. University of California Press, Berkeley.

Goldberg, D. (1989). *Genetic Algorithms in Search, Optimization, and Machine Learning*. Addison-Wesley, Reading, MA.

Gould, S. J. and Lewontin, R. (1979). The spandrels of san marco and the panglossian paradigm: A critique of the adaptationist programme. *Proceedings of the Royal Society*, B205:581–598.

Holland, J. (1975). *Adaptation in Natural and Artificial Systems*. University of Michigan, Ann Arbor.

[54]The original version of the paper had a long section of objections and replies, which I have dropped for reasons of space and because some of the entries are now included in Nickles, 2003a.

Holland, J. (1992). Genetic algorithms. *Scientific American*, pages 66–72.

Holland, J. (1995). *Hidden Order: How Adaptation Builds Complexity*. Addison-Wesley, Reading, Mass.

Hull, D. (1988). *Science as a Process: An Evolutionary Account of the Social and Conceptual Development of Science*. University of Chicago Press, Chicago.

Johnson-Laird, P. and Byrne, R. (1991). *Deduction*. Lawrence Erlbaum, Hillsdale, N.J.

Kantorovich, A. (1993). *Scientific Discovery: Logic and Tinkering*. SUNY Press, Buffalo.

Kelly, K. (1994). *Out of Control: The New Biology of Machines, Social Systems, and the Economic World*. Addison-Wesley, Reading, Mass.

Kimura, M. (1983). *The Neutral Theory of Molecular Evolution*. Cambridge University Press, Cambridge.

Koza, J. (1992). *Genetic Programming: On the Programming of Computers by Means of Natural Selection, vol. I*. MIT Press, Cambridge, Mass.

Koza, J. (1994). *Genetic Programming II: Automatic Discovery of Reusable Programs*. MIT Press, Cambridge, Mass.

Koza, J., Bennett III, F., Andre, D., and Keane, M. (1999). *Genetic Programming III: Darwinian Invention and Problem Solving*. Morgan Kaufmann, San Francisco.

Koza, J., Keane, M., Streeter, M., Mydlowec, W., Yu, J., and Lanza, G. (2003). *Genetic Programming IV: Routine Human-Competitive Machine Intelligence*. Kluwer, Dordrecht.

Koza, J. R., Yu, J., Keane, M. A., and Mydlowec, W. (2000). Evolution of a controller with a free variable using genetic programming. Volume 1802 of *Lecture Notes in Computer Science*, pages 91–105. Springer-Verlag, Berlin.

Kuhn, T. (1962). *The Structure of Scientific Revolutions*. University of Chicago Press, Chicago.

Kuhn, T. (1978). *Black-Body Theory and the Quantum Discontinuity, 1894-1912*. Oxford University Press, Oxford.

Langley, P., Simon, H., Bradshaw, G., and Zytkow, J. (1987). *Scientific Discovery: Computational Explorations of the Creative Processes*. MIT Press, Cambridge.

Laudan, L. (1980). Why was the logic of discovery abandoned? In Nickles, T., editor, *Scientific Discovery, Logic, and Rationality*, pages 173–183. Reidel, Dordrecht. Reprinted in Laudan Laudan, 1981, pages 181-91.

Laudan, L. (1981). *Science and Hypothesis*. Kluwer, Dordrecht.

Louis, S. Website: http://www.cse.unr.edu/ sushil/.

Margolis, H. (1987). *Patterns, Thinking, and Cognition: A Theory of Judgment*. University of Chicago Press, Chicago.

Mitchell, M. (1996). *An Introduction to Genetic Algorithms*. MIT Press, Cambridge, Mass.

Mitchell, M. and Forrest, S. (1995). Genetic algorithms and artificial life. In Langton, C., editor, *Artificial Life: An Overview*. MIT Press, Cambridge, Mass.

Nagel, T. (1976). Moral luck. Reprinted in his *Mortal Questions*. Cambridge University Press, Cambridge, 1979, chap. 3, and in Statman (1993), pages 57-75.

Newell, A. and Simon, H. (1972). *Human Problem Solving*. Prentice-Hall, Englewood Cliffs, NJ.

Nickles, T. (1981). What is a problem that we may solve it? *Synthese*, 47:85–118.

Nickles, T. (1984). Positive science and discoverability. *PSA 1984*, 1. Edited by P. Asquith and P. Kitcher. East Lansing: Philosophy of Science Association:13–27.

Nickles, T. (1985). Beyond divorce: Current status of the discovery debate. *Philosophy of Science*, 52:177–206.

Nickles, T. (1987). From natural philosophy to metaphilosophy of science. In Kargon, R. and Achinstein, P., editors, *Kelvin's Baltimore Lectures and Modern Theoretical Physics: Historical and Philosophical Perspectives*, pages 507–541. MIT Press, Cambridge.

Nickles, T. (1992a). Epistemic amplification: Toward a bootstrap methodology of science. In Brzezinski, J., Coniglione, F., and Marek, T., editors, *Science: Between Algorithm and Creativity*, pages 29–52. Eburon, Delft.

Nickles, T. (1992b). Good science as bad history: From order of knowing to order of being. In McMullin, E., editor, *The Social Dimensions of Science*, pages 85–129. University of Notre Dame Press, Notre Dame.

Nickles, T. (1995). History of science and philosophy of science. *Osiris*, 10:139–163.

Nickles, T. (2003a). Evolutionary models of innovation and the Meno problem. In Shavinina, L., editor, *International Handbook of Creativity, vol. 1*, pages 54–78. Elsevier Scientific Publications, Amsterdam.

Nickles, T. (2003b). Normal science: From logic of science to case-based and model-based reasoning. In Nickles, T., editor, *Thomas Kuhn*, pages 142–77. Cambridge University Press, Cambridge.

Olby, R. (1979). Mendel no mendelian? *History of Science*, 17:53–72.

Peirce, C. S. (1877). The fixation of belief. Reprinted with changes in *Collected Papers of Charles Sanders Peirce*, vol. 5. Edited by C. Hartshorne and P. Weiss. Harvard University Press, Cambridge, Mass., 1934, pages 358–387.

Pickering, A. (1984a). Against putting the phenomena first: The discovery of the weak neutral current. *Studies in History and Philosophy of Science*, 15:85–117.

Pickering, A. (1984b). *Constructing Quarks: A Sociological History of Particle Physics*. University of Chicago Press, Chicago.

Plotkin, H. (1993). *Darwin Machines and the Nature of Knowledge*. Harvard University Press, Cambridge, Mass.

Popper, K. (1934). *Logik der Forschung*. Hutchinson, London.

Popper, K. (1963). *Conjectures and Refutations*. Routledge and Kegan, London.

Popper, K. (1972). *Objective Knowledge: An Evolutionary Approach*. Clarendon Press, Oxford.

Popper, K. (1974). Response to Campbell. In Schilpp, 1980, pages 1059–1065.

Reitman, W. (1964). Heuristics, decision procedures, open constraints, and the structure of ill-defined problems. In Shelly, M. W. and Bryan, G. L., editors, *Human Judgments and Optimality*, pages 282–315. John Wiley, New York.

Rescher, N. (1990). Luck. *Proceedings of the American Philosophical Association*, 64:5–19.

Rheinberger, H.-J. (1997). *Toward a History of Epistemic Things: Synthesizing Proteins in the Test Tube*. Stanford University Press, Stanford.

Schaffer, S. (1986). Scientific discoveries and the end of natural philosophy. *Social Studies of Science*, 16:387–420.

Schaffer, S. (1994). Making up discovery. In Boden, M., editor, *Dimensions of Creativity*, pages 13–51. MIT Press, Cambridge, MA.

Schaffner, K. (1980). *Discovery and Explanation in Biology and Medicine*. University of Chicago Press, Chicago.

Schilpp, P. A., editor (1980). *The Philosophy of Karl Popper*. Open Court, LaSalle, Ill.

Schuster, J. (1977). *Descartes and the Scientific Revolution: 1618-1634*. UMI Dissertation Services, Ann Arbor.

Schuster, J. and Yeo, R., editors (1986). *The Politics and Rhetoric of Scientific Method*. Reidel, Dordrecht.

Sellars, W. (1956). Empiricism and the philosophy of mind. In Feigl, H. and Scriven, M., editors, *Minnesota Studies in the Philosophy of Science, vol. 1*, pages 253–329. University of Minnesota Press, Minneapolis.

Simon, H. (1973). The structure of ill structured problems. *Artificial Intelligence*, 4:181–201. Reprinted in *Models of Discovery*, Reidel, Dordrecht, 1977, pages 304-325.

Simon, H. and Lea, G. (1990). Problem solving and rule induction: A unified view. In Shavlik, J. and Dietterich, T., editors, *Readings in Machine Learning*, pages 26–43. Morgan Kaufmann, San Mateo.

Simonton, D. (1988). *Scientific Genius: A Psychology of Science*. Cambridge University Press, Cambridge.

Solomon, R. (1981). *Introducing the German Idealists*. Hackett, Indianapolis.

Souriau, P. (1881). *Theorie de l'Invention*. Hatchette, Paris.

Statman, D., editor (1993). *Moral Luck*. SUNY Press, Albany.

Toulmin, S. (1972). *Human Understanding*. Princeton University Press, Princeton.

Wiener, N. (1948). *Cybernetics: Or, Control and Communication in the Animal and the Machine*. Wiley, New York.

Williams, B. (1976). Moral luck. Reprinted in his *Moral Luck*. Cambridge: Cambridge University Press, 1981, chap. 2, and in Statman (1993), pages 35-55.

Winston, P. H. (1992). *Artificial Intelligence*. Addison-Wesley, Reading, Mass. Third edition.

Wolpert, D. (1996). The lack of a priori distinctions between learning algorithms. *Neural Computation*, 8:1341–1390.

Wolpert, D. and Macready, W. (1997). No free lunch theorems for optimization. *IEEE Transactions on Evolutionary Computation, vol. 1*, pages 67–82. Condensed version of 1995 Santa Fe Institute Technical Report, SFITR 95-02-010, "No Free Lunch Theorems for Search".

Wright, L. (1973). Functions. *Philosophical Review*, 82:139–168.

TRADITION AND INNOVATION: EXPLORING AND TRANSFORMING CONCEPTUAL STRUCTURES

Matti Sintonen
Department of Philosophy
University of Helsinki
matti.sintonen@helsinki.fi

1. Introduction

On one attractive view creativity is exploration of conceptual structures or spaces, defined by a descriptive vocabulary and a grammar, and generative rules for producing admissible outcomes (Boden, 1994b; Langley et al., 1987). To be creative is to explore and perhaps transform a conceptual space or conceptual structure. Discoveries, scientific, artistic, practical or any other are simply results accumulated in a creative enterprise. Although one does not always need special mental qualities to make discoveries, ideas that make an impact and therefore really merit the title are unlikely to arise without open-minded exploration of the boundaries of the conceptual spaces. And although serendipity exists and luck helps, novelties tend to frequent flexible and persistent explorers.

Let us call this the explore-and-transform-paradigm of creativity and discovery. It seems to be winning the day, and rightly so, especially because it makes it plain that creativity presupposes rather than opposes tradition. There is, however, something which could be called Boden's problem. Although conceptual structures are crucial for "the identification and evaluation of creativity", there is, as Boden puts it, no simple or uniform measure for their "depth" ascross all types of creative activity.

This paper takes a look at the credentials of the paradigm in opposing quarters of the scientific enterprise. The suggestion is that both the nature of the activity and the virtues of creativity are different in different areas. First, some applied types of inquiry do not have enough of elaborate and well-defined structures for the explore-and-transform-paradigm to work. These messy fields are increasigly important nevertheless. To avoid downgrading their worth within the enterprise of knowledge one could call them intractable fields, following

J. Meheus and T. Nickles (eds.), *Models of Discovery and Creativity*,
DOI 10.1007/978-90-481-3421-2_10, © Springer Science + Business Media B.V. 2009

the suggestion of the ecologist Slobodkin (1988) who has managed to thrive in the midst of a cognitive (and administrative) chaos.

The second difficulty arises in fields where there are such structures, namely the highly mature fields with hierarchical and layered structures. This is the proper domain of the explore-and-transform-paradigm of creativity, for here there is an established understanding as to what the problems are and how they are to be dealt with. The difficulty here is that the structures move in two dimensions, synchronic and diachronic, and this presents problems for discovery. Very briefly, discoveries have socio-historical depth which makes the identification of the crucial explorers and transformers—the discoverers—difficult. To sum up, in the first case problems do not have enough depth for the explore-and-transform-paradigm of creativity to work, in the second case they have too much structure for a simple notion of discovery. Together these difficulties question the plausibility of a unitary notion of creativity and discovery.

I shall start by contrasting tradition and innovation (§2) and by suggesting that the so-called semantic view of theories improves our understanding of creativity and discovery for it allows hierarchical models in concept formation and theory testing (§3). Coupled with recent advances in cognitive science it is more promising still. I shall then take a look at the challenge of intractable fields (§4) and move to the more tractable ones. Here we face the fundamental complaint of the historicists (§5), viz., that structures built by logical-reconstructionist tools, whether sets of sentences or models described in the current language of science, viz. mathematical English, make no explicit room for historical depth. But I'll suggest (§6) that we need the further step in which scientific theories are conceived as evolving (and occasionally revolving) individuals, with historical trajectories as well as "snap-shot" structures.

2. Taditionalists and Iconoclasts

But what are creativity and innovativeness? Scientists have a professional interest in questions that improve on common sense ("ordinary language") questions in being more discerning and answerable. Creativity, then, amounts to phrasing and answering ever more refined questions made possible by concepts and laws and models of a scientific theory (and in other fields by the conventions of an art, possibly encoded in a system of representation such as musical notation, or the generative apparatus of any like activity). And the person who manages to refine the generative apparatus by altering one or more of the generative rules is more creative still. Finally, the real giants in the sciences or in the arts are individuals who, having mastered the conventions, make a successful case for replacing them by another one. In the terminology adopted this means opening ever new areas of questions for scrutiny. The implications of this view to various arts and sciences might well be different, for the structures mapped by the rules and conventions might serve different ulterior purposes, such as truth and fun.

But why should finding something new be such a big deal? The idea that one should aim at novelties fits the modern emphasis on individuality and the value of seeing or making something completely unprecedented. For a thinker in the antiquity or the middle ages such an idea would have been preposterous, unfounded bolstering of one's ego over others. The notion that one could bring about something that was both novel and valuable without relying on a collectively sustained tradition which both makes it possible to see novelties and sets the boundaries for their nature was nearly unintelligible. The reason is of course that to be worth making, a novelty must relate to the concerns of the community—and these concerns were encapsulated in the tradition.

This brings us back to the title of this paper, and to its problems. Problem number one is a workable characterization for novelty. Problem number two is the built-in tension between tradition and innovation in any such characterization. In a sense the new historiography marks a return to the earlier emphasis on tradition. The phrase tradition and innovation in the title refers to a seminal talk which Thomas Kuhn held almost forty years ago, entitled *The Essential Tension: Tradition and Innovation in Scientific Research.* It was delivered in a conference on the identification of scientific talent, and the title later became the title of a collection of essays. The basic idea of the paper was simple. As Kuhn pointed out, many head hunters looking for scientific (or artistic, or any other) talent are advised to look for imaginative and divergent minds. However, he said, at least in mature natural sciences intellectual flexibility and divergent thinking is not enough. True, to the extent that breakthroughs are not results of mere applications—and this seems to hold almost by definition—open mindedness is a must. But scientific revolutions are vanishing episodes in normal scientific puzzle solving. I shall quote:

> Almost none of the research undertaken by even the greatest scientists is assigned to be revolutionary, and very little of it has any such effect. On the contrary, normal research, even the best of it, is a highly convergent activity based firmly upon a settled consensus acquired from scientific education and reinforced by subsequent life in the profession ... As I shall indicate below, only investigations firmly rooted in the contemporary scientific tradition are likely to break that tradition and give rise to a new one. (Kuhn, 1977, p. 140)

And Kuhn concluded the paragraph by saying that very "often the successful scientist must simultaneously display the characteristics of the traditionalist and of the iconoclast".

3. Scientific Structures

How about creativity in science. Could there be a metatheoretic notion which could back up our verdicts on creativity and discovery, to give heroes and ordinary scientists their due? With the abundance of creativity folklore the answer seems to be self-evidently positive. And after all, however much rival philosophies of science vary in detail and emphasis, all seem to agree that scientific theories simply *are* conceptual structures, hooked onto the world (here and there) through rules for semantic interpretation. To learn the trade is to learn these structures, and to be creative is to produce applications new to the individual scientist (Boden's P-creativity) or to the scientific community (H-creativity). For instance, on the now discredited positivist view theories are sets of sentences with well-defined structures laid down by the theoretical postulates and the rules for semantic interpretation.

For several reasons, no longer worth detailing, the received view of theory structure is a non-starter. But its most serious modern rival is worth considering, not because it denies that theories are deductively organized sets of sentences, but because it represents theories in terms of various types of families of models. This is the so-called semantic view of theories. In the state space approach, scientific theories are formalized as classes of models. These models are formed from variables which define the state space and equations whose solutions present the possible transformations of the states of the given state space (laws of succession) or indicate the possible states of the system (laws of coexistence). The structure of scientific theories, according to the state space approach, is defined by specifying models of the theory directly, without recourse to the possible axiomatizations of the theory (Van Fraassen, 1970).

I think the semantic view is an important step towards elaborating the explore-and-transform-paradigm because it explicitly recognizes that sciences are full of vertically and horizontally arranged structures. Elizabeth Lloyd (1988) has shown how theory confirmation in biology proceeds along structured paths. According to Lloyd (1988, pp. 244–245), describing the structure of a theory only involves specifying the set of models of the theory. More specifically, constructing a model within a theory involves locating an idealized system of the sort defined by the theory in the state space of the theory. Lloyd follows Patrick Suppes's idea of a hierarchy of theories through which natural systems and ideal systems are linked. First, there are theoretical models which can be employed to make empirical claims. However, theoretical models are too abstract and must be specified (or concretisized) to experimental models concerning specific types of experiments. Finally, there are models of data which anchor experimental models to specific performances of an experiment, where possible realizations of data are delimited. Thus a theoretical model is brought, step by step, down to earth for direct assessment to be possible.

Next, take Deborah's Mayo's (1996) views on the experimental inquiry. Although Mayo does not endorse the semantic view as such her account of inquiry as moving along a series of experimental models is best viewed in its light. On her error theorist approach, experimental inquiry proceeds along series of models which she calls primary models, experimental models, and data models. There always is a framework of inquiry with methods organized around the hierarchy of these models, and the methodological rules are more like strategic advice for promoting the goal of severe testing and hence, for learning from errors. The interesting thing about such a model is that it gives a refined and hierarchical account of theory testing. Testing is not a one-step confrontation between a hypothesis and evidential sentences but a step-by-step procedure for "building up, correcting, and filling out the models needed for substantiating severe tests". Lloyd's and Mayo's accounts are interesting also because they extend creative exploration into testing and confirmation, thus in effect breaking the boundary between the contexts of justification and inquiry.

The next step towards understanding creativity comes from cognitive science. The received view provides no account, or at best a very weak account, of what concepts are and they say next to nothing about how conceptual networks change. But as Paul Thagard (1992, p. 30) observes, "an understanding of conceptual revolutions requires much more than a view of the nature of isolated concepts. We need to see how concepts can fit together into *conceptual systems* and what is involved in the replacement of such systems". This is precisely what his suggestion does. As a result we have conceptual systems in which concepts are organized into hierarchies by help of various types of links, such as kind, instance, rule, property and part links. To the extent scientific conceptual systems consist of networks of nodes and links analogous to other cognitive systems or structures, we have an account of conceptual change. As Thagard puts it, it is adding or deleting nodes and links, and adding or deleting rules. From the point of view of understanding creativity and discovery the most important feature of Thagard's notion of conceptual change is that it uses tools which have been rooted in well-established research programmes in AI, cognitive psychology and cognitive science. Furthermore, it has the crucial advantage that it provides perhaps the most plausible account available for the importance or significance of a conceptual change, whether adding or substracting or altering (which might be viewed as substraction followed by addition). Finally, it enjoys extra appeal because it gives both tradition and innovation their due. Conceptual revolutions involve, by definition as it were, dramatic replacing of major portions of the conceptual systems. Nevertheless continuity prevails because some of the links to other concepts are retained. Dramatic changes are seen in hierarchies of concepts built of what Thagard calls kind links and part links. They provide a framework within which concepts are arranged and organized. As Thagard sums up, "changes in kind-relations and part-relations usually involve a restructuring of conceptual systems that is qualitatively different from mere addition or deletion of nodes and links" (Thagard 1992, p. 32).

4. Applied and Intractable Fields

These advances are important steps towards a better understanding of scientific structures—and the list could be extended. However, Boden's problem (1994b) does not disappear: despite these advances there is no simple measure for the "depth" of the relevant dimensions in which creative achievements might be assessed. Although particular ideas can be compared (e.g., within the Frank Lloyd Wright "grammar" for prairie houses, a two-fireplace house would be more daring and creative than one with an added balcony), there is no general solution. The most serious difficulty is that much of science is not carried out under well-crystallized conceptual schemes or frameworks or patterns. Applied sciences, to the extent that they form a unitary group of sciences at all, do not happily fit into this picture.

To set the stage it is worth asking what is scientific knowledge and why is it worth having? Scientific knowledge differs from ordinary or common sense knowledge in being more systematic, empirically (observationally and experimentally) controlled, more truthlike, simple and consilient, progressive and what have you. But to condense these aims to a short *precis*, science seems to have two types of aims, theoretical understanding and practical utility. And scientific viewpoints, research programs and especially theories with generalizations and nomic laws are valued to the extent they are carriers of these cognitive objectives or desiderata. Behind theoretical understanding, in turn, seems to be some combination of cognitive values such as truth, information content, and explanatory power, as well as such "aesthetic" virtues as simplicity and conceptual economy, i.e., the mastery of a wide variety of areas by a minimum of conceptual apparatus. To give a working definition, let us then say that basic research is search of knowledge for its own sake, and thus aims at the maximization of cognitive desiderata.

I am not in fact suggesting that the distinction between applied and basic inquiry can be drawn within academic departments and research institutions, nor that even disciplines or fields could be allotted to these categories. I therefore accept that projects within basic sciences can have clearly applied aspirations and vice versa. But I do suggest that there are more locally inspired projects which do not exhibit similar profiles of inquiry. Shall we say, tentatively, that there are theory-driven, experiment-driven and application-driven or applied fields. Theory-driven domains contain highly mathematical fundamental theories which more or less specify what counts as a research question—and what counts as an answer. Applied fields do not have one or two overarching theories capable of delineating questions—rather, problems are problems because they have a closer tie to practical interest. As a result, these fields are populated by rival approaches.

Applied research is pursuit of knowledge where the goal is, to employ OECD's authoritative characterization from some 30 years ago, to attempt "to put to use the findings of basic research or even to discover new knowledge which might have immediate practical application".[1] Here applications are singled out by other than purely cognitive goals. The deterioration of the environment can be considered a problem which is, to put it mildly, not a merely academic one. Consequently, promoting the well-being of the environment gives a practical goal and hence an extra-scientific criterion of relevance for what counts as an answer or a good answer: goodness depends on how well the answer serves to promote the practical desideratum.

As an example of the occupational hazards in an intractable field, let us take research conducted in fishing farming. Salmon has become a major industry with ever growing economic potential. In Norway it resulted in applied research directed at maximizing output, with tragic results, such as spreading of diseases and infections and of massive use of antibiotics, not to speak of pollution of rivers and seas used for spawning and breeding (see Kaiser, 1993). Even gathering of basic data on salmon met with obstacles because there was no theory which said when salmon where likely to escape from fish yards, or what happened to wild salmon. And when some of the problems were tackled it turned out that the expertise of veterinarians was often useless because, to put it simplistically, cows are cows and salmon are salmon. Basically of course knowledge from fundamental ecological theories could be of help, but the path from widely applicable ecological theories to realistic models for the behavior of fish in the rather contrived conditions—sometimes with the equivalent of 8–10 salmon swimming in a bath tub—is highly tortuous. In any case it is a far cry from the neatly hierarchical systems of knowledge found in mature fields.

It might be suggested that theoretical and applied ecology are still immature and simply waiting for the right unitary point of view and conceptual apparatus. But especially in areas where non-biological viewpoints are crucial, in resource management and conservation biology for instance, too many concerns are involved to make this response feasible. An alternative view is that ecology does not focus on a couple of theoretically central (and "feasible") problems at a time. As Slobodkin (1988, p. 338) puts it, "ecology may be the most intractable of all legitimate sciences ever developed".

[1] This characterization is of course loaded with interpretative issues concerning "immediate", "application", and "practical". It would not go to insist that basic science does not focus on applications, for it is a fundamental requirement of any descriptive and explanatory network of concepts that it can be interpreted and thus applied. However, intended applications in basic science arise from the purely descriptive and explanatory perspective of telling how things are and why they are the way they are.

I now come to the challenge applied research and intractable fields provide for accounts of creativity and discovery. Where the big initial questions are motivated through practical needs it is natural that not all potential answers are tied to this or that particular theory, approach or viewpoint. A big question does not as such specify what conceptual equipment the answer should have. This means that much of applied research is theoretically eclectic: although the generation of potential answers involves focusing on a particular approach, rivals are abundant, and no approach is a priori illegitimate (see Sintonen, 1990). Theoretical eclecticism, in turn, brings different virtues to the foreground. Interestingly enough Kuhn writes:

> The problems among which they may choose are likely to be largely determined by social, economic, or military circumstances external to the sciences ... It is, I think, by no means clear that the personality characteristics requisite for pre-eminence in this more immediately practical sort of work are altogether the same as those required for a great achievement in basic science. (Kuhn, 1977, p. 238)

What these types of inquiry have in common is, precisely, lack of a unitary point of view or disciplinary matrix, to use Kuhn's apt phrase. In such areas the problem is first of all to decide what the problem is, and then to make it more precise. The cognitive profiles of the cooperating disciplines, the mother science so to speak, can be alien to each other to differing degrees.

Leo Apostel and others distinguish between several kinds of relationships between cooperative disciplines, ranging from multidisciplinary research in which conceptually alien disciplines (musicology, ethnology, philology) approach a common problem to pluridisciplinary research (traditionally close disciplines engaged in a joint venture), and interdisciplinary research in which participants, trained within different fields or disciplines, with their own concepts, methods, generalizations, theories, embark on a coordinated effort to solve a problem or set of problems (see Apostel et al., 1972).

Here we need flexibility and willingness to try alternatives, ability to adopt differing conceptual grids, communication and negotiation skills, for a pluridisciplinary approach presupposes ability to transgress one's disciplinary boundaries. It does not mean that such research is completely at odds with the explore-and-transform-paradigm, but it does mean that being creative presupposes the ability to work with a multitude of conceptual structures at one and the same time.

5. Discovery in the Mature Sciences

Now, let us leave the difficulties of applied sciences and intractable fields and ask if the explore-and-transform-paradigm of creativity and discovery, with its kit filled with essentially psychological and cognition-scientific tools can be applied in the more clearly basic pursuits, especially mature fields. The answer is divided. It seems to work for creativity for indeed exploring structures is a recognized recipe for success. However, it needs some reworking before it can illustrate scientific discovery, mainly because discoveries are in part historical and social reconstructions. I think it would be unwise to rule the paradigm out completely, for reasons to be made clear shortly. But there is a difficulty and a puzzle. The difficulty is that of reconciling the forward-looking engineering point of view of cognitive science with the backwards-looking point of view of history of science.

And there is a genuine puzzle here too, viz., the question whether scientists *qua* scientists need history. For if the engineering point of view is correct, a scientist (or a group of scientists, although this might alter the problem) at any given time faces the current theories, problems and tools for solving them. What is buried in the past can influence current decisions only indirectly. By definition, what is relevant is here and now. So why should she or he bother with the opinions, true or false, fruitful or barren, of previous generations working in the field? It might be interesting to know, and it might enhance one's historical sense and therefore self-understanding, but would it make any difference to the task if he adopted a stubbornly contemporary point of view?

The historicists have suggested, since R. G. Collingwood and even earlier, that there is always more to a research question than meets the untutored ear. Scientific claims are answers to structured but opaque questions—and the structures are inherited from the background theories and constraints accepted at a given time. One-dimensional logical reconstructions given in syntactic and semantic terms identify scientific questions and answers, such as explananda, explanantia, discoveries, and scientific theories, with surface descriptions (Newton's theory of gravitation, Mendel's theory of heredity) and thereby miss the deep structures and "local" contexts. But why would that be a problem for the explore-and-transform-paradigm? It creates a problem for the identity of the results of scientific talent or genius, viz. scientific discoveries. One and the same "phenomenon" can find a niche in different conceptual structures. A description in the mouth of the classical physicist does not mean the same, in an important sense, than it does in the mouth of the relativist. And if this is the case, there is no such thing as *the* discovery of a phenomenon or theory, independently from this or that conceptual structure. Consequently, not only are discoveries possible only by building on an already available tradition but the very idea of finding who discovered what becomes a suspicious one. The only alternative seemed to be to give up logical-reconstructionism in favour of historicism.

The difficulties have been widely recognized and they no doubt in part lead to the new historiography which takes the social and cultural context of discovery seriously. Early proponents of this view included, e.g. Duhem and Bachelard, Ludwig Fleck, and in the analytic tradition Kuhn, Toulmin and Feyerabend, just to name a few. If surface descriptions of scientific discoveries and creations can be embedded in several conceptual structures it may even be difficult, at a frontier of a field, to distinguish between the various issues. It is possible, then, that there is talking past one another, semantic incommensurability or practical blurring of boundaries.[2] The challenge to attempts to provide a measure of conceptual complexity and depth for scientific problems, needed for the explore-and-transform-paradigm of discovery, is that the logical-reconstructionist view is constitutionally incapable of taking into account historical depth. This is what Stephen Toulmin argued when he suggested that theories are historical entities with an evolutionary past.

6. Exploring Paradigms

Is there any way to accommodate the undeniable historical dimensions within an essentially structured notion of a theory? The semantic view discussed earlier represents theories by help of families of models but has an underdeveloped sense of diachrony. Fortunately its continental cousin, the so-called structuralist approach, improves on this by giving theories a snap-shot "grammar" and by embracing an explicitly historical perspective. I shall suggest, following the historicists and contemporary structuralists in philosophy of science, that the word "theory" can refer to an entire net of interrelated theory-elements which, especially in mature fields, have internal elaborate snap-shot structures. And most importantly, since theory-nets are historically evolving and revolving entities, they are ideal for representing growth of knowledge. It is no accident that Kuhn's comment on the proposal was overwhelmingly positive. He wrote that the view is the first formal or semiformal explicate which comes anywhere near representing his views of growth of knowledge.

I shall skip all formal details and only present the fundamental ideas needed to appreciate its virtues in explicating creativity and discovery. On this view a theory-element T is an ordered pair $\langle K(T), I(T)\rangle$ in which $I(T)$ is the set of intended applications of the theory and the $K(T)$ the theory-core, more precisely a quintuple $\langle M_{pp}(T),\ M_p(T),\ M(T),\ GC(T),\ GL(T)\rangle$. The most elementary structural units of theories are its models $M(T)$ (the laws of the theory), i.e. sequences of basic sets and relations over some of these sets. The basic sets $D_1 \ldots D_m$ give the theory's ontology by specifying the real or empirical and mathematical objects needed, while the relations $R_1 \ldots R_m$ are (usually quantitative) functions from these objects to real numbers (or vectors). Apart from models there are the theory's potential partial models $M_{pp}(T)$, structures of which it makes sense to ask whether they can be enriched with theoretical

[2] It may well be that there is a difference between the sciences and the arts here, for playing with the conventions of different traditions is a valued means of achieving effect in the arts. However, I cannot go into these issues here.

functions so as to satisfy the laws M of the theory, and $M_p(T)$, the potential models which do include the theoretical functions.

These two types of structures are the most important ones. Intuitively speaking they contain the distinction between "frame conditions" and "substantial laws", those structures which are of the right type, and those which satisfy the substantial axioms of the theory. Such a way of conceiving the identity of the core of a theory-element already gives a relatively rich structure, for one can explore (and transform) laws without touching the frame conditions. But theories in mature fields are not solitary individuals. The models of a theory characteristically contain relationships between one another as well as links to other theories, represented by the global constraint $GC(T)$ and the global link $GL(T)$.

However, single theory-elements do not suffice to describe all aspects of question-answer dynamics. Theories do not hatch as finished products sufficient to deal with all forthcoming applications. Rather, they are conceived in the form of gappy structures which must be nurtured until they turn into powerful theories. Theory-elements in science characteristically conspire to form theory-nets N, sequences of theory-elements $T_1, T_2, \ldots, T_n$ connected with one another by the specialization relation, and theory-holons H, still larger entities comprising theory-elements from different theory-nets (Moulines, 1996). A theory-net in turn has one or more basic theory-elements $B(N)$ and a number of specialized theory-elements $\langle K_i, I_i \rangle \in N$, introduced to make more specific claims about some more limited classes of applications ($I_i \subseteq I_0$). The basic core of the theory-net may then give rise to several branches of specializations, and the result may be a hierarchial tree-structure.

Theory-nets, in turn, are historically evolving individuals. It follows that theories have both horizontal and vertical structure, and complex links to elements in other fields (see Balzer and Moulines, 1996). When a theory is proposed it usually contains few well-motivated applications. The understanding is that later generations refine and expand the theory-net to cover the remaining envisioned but so far unexamined or unsuccessfully examined applications. A theory-evolution represents such historical development: it is a finite sequence of $\langle K_0, I_0 \rangle$-based theory-nets $N_1, N_2, \ldots$ such that each N_{i+1} contains at least one theory-element obtained by specialization from an element in the historically preceding theory-net N_i.

It is now easy to appreciate the implications for creativity and discovery, in outline. The frame and substantial assumptions, together with the other elements of the theory-element provide structures which make it possible to express some claims and questions, but not others. A scientific community, the holders or supporters of a theory during a certain historical period, in fact subscribe to a theory-net and its elements which literally propose a host of more or less well-defined questions, including yes-no-questions, concerning systems on which the community focuses during the period. There are also clarification, precization and classification of questions, requiring refinement of terms or values or classification of phenomena to classes of intended applications as answers. There are important questions concerning values for constants, and why-questions requiring explanatory answers.

Although the structuralist view does not solve all problems its virtues are worth spelling out. First, since theory-nets literally feed in empirical questions concerning structures (what special laws and theoretizations do we need to say more about such and such structures) they in fact produce a model for exploring the resources of the paradigm. There is also a finesse in the structuralist proposal which requires special mention, viz, the role of intended applications. Theories are identified through their cores and sets of intended applications. Formally, intended applications are subsets of M_p's or M_{pp}'s, thus guaranteeing that they have the right structure. Inquiry, then, proceeds by refining questions by finding suitable vocabularies to express answers, and by exploring possible laws expressed by help of these vocabularies. When joined with the idea that search for specializations and theoretizations involves strategic thinking, the model refines the familiar notion that asking a good question at the right time is more effective than performing countless but aimless deductions. My suggestion is that by thus focusing on intended applications and the locally available tools in a tradition we get an inspiring notion of a strong heuristic.

This view has several advantages. Since theories have both instantaneous structure and historical depth the view is particularly suitable for the purposes of history of science and science studies. It makes it possible to see why it usually is impossible to put a precise date to a discovery: since discoveries cannot be identified with their surface representations, scientists equipped with different background theories and methodological and substantial assumptions may in fact have had different problems in mind. This would also explain why the so called multiple discoveries should not be taken at face value (for this, see Schaffer, 1996).

The metatheory can also explicate the notions of importance, centrality or relevance so important for assessing discovery and creativity. Clearly, in hierarchically organized conceptual networks the fundamental laws of a theory net are more important than the special laws designed for more restricted applications. Furthermore, the structuralist theory notion contains the notions of inter-theory relations and links which generate a structure within the elements of a theory net. Similarly, there is in the structuralist view a global structure to science at large, indicating that one element can interpret another element even in a completely different theory net.

Now, such a structure does not necessarily specify a unique order within the concepts, laws, and links of a theory-net. Nevertheless it does give a working notion of centrality. Secondly, because structuralism views theory-nets as evolving entities with basic theory elements as there cores there is the possibility of explicating historical centrality or importance. It goes without saying that there can be no unique ahistorical notion for this intuitive concept, because, as was the case with other theory notions, concepts and special laws can be embedded in different theory-nets.

References

Apostel, L., Berger, G., Briggs, A., and Michaud, G., editors (1972). *Interdisciplinarity. Problems of Teaching and Research in Universities*. Centre for Educational Research and Innovation, Paris.

Balzer, W. and Moulines, C. (1996). *Structuralist Theory of Science. Focal Issues, New Results*. de Gruyter, Berlin and New York.

Boden, M., editor (1994a). *Dimensions of Creativity*. MIT Press, Cambridge, Ma, and London.

Boden, M. (1994b). What is creativity. In Boden, 1994a, pages 75–117.

Kaiser, M. (1993). Some thoughts on the responsibility of scientists in relation to the growth of fish farming in norway. In Welin, S., editor, *Studies in Research Ethics, 2: Scientific Responsibility and Public Control*, Göteborg.

Kuhn, T. (1977). *The Essential Tension*. University of Chicago Press, Chicago.

Langley, P., Simon, H., Bradshaw, G., and Zytkow, J. (1987). *Scientific Discovery. Computational Explorations of the Creative Processes*. MIT Press, Cambridge, Massachusetts and London.

Lloyd, E. (1988). *The Structure and Confirmation of Evolutionary Theory*. Greenwood Press, Wesport.

Mayo, D. (1996). *Error and the Growth of Experimental Knowledge*. The University of Chicago Press, Chicago and London.

Moulines, U. C. (1996). Structuralism: The basic ideas. In Balzer, W. and Ulises Moulines, C., editors, *Structuralist Theory of Science*, pages 1–13. de Gruyter, Berlin and New York.

Schaffer, S. (1996). Making up discovery. In Boden, 1994a, pages 13–51.

Sintonen, M. (1990). Basic and applied research: Can the distinction still be drawn? *Science Studies*, 2:23–31.

Slobodkin, L. (1988). Intellectual problems of applied ecology. *Bioscience*, 38:337–342.

Thagard, P. (1992). *Conceptual Revolutions*. Princeton University Press, Princeton and New Jersey.

Van Fraassen, B. (1970). On the extension of Beth's semantics of physical theories. *Philosophy of Science*, 37:325–338.

A PURPOSEFUL ALLIANCE IN THE SERVICE OF CREATIVE RESEARCH

The Network of Vitamin Investigators

Petra Werner
Berlin-Brandenburgian Academy of Science, Berlin

1. Introduction

Many items of scientific knowledge are recognized, both by historians of science and the general public, to be collective scientific achievements. The knowledge in question is "collective work" in the sense that it results from discoveries and inventions that are continuously interconnected and, although independent of one another, also overlapping one another.

In the twentieth century, however, it has happened frequently in fields such as the investigation of natural substances, and biochemistry, that different investigators could attain credit for simultaneously recognized results, a situation which was expressed through the award of Nobel Prizes to several investigators for the same theme.

International competition and the simultaneity of individual discoveries and their intermediate steps led to very sharp conflicts, which are known to us in many fields of investigation. These conflicts can arise from various sources. If one looks at basic research, then it is essentially the desire for recognition "for the first formal presentation of an innovation or discovery to the scientific community" (Hagstrom, 1965, p. 69). Priority is the equivalent of currency in science—sometimes, if the discovery has commercial applications, it can even (as in the cases I have considered), be translated literally into cash.

Robert Merton, as well as Hagstrom, undertook investigations during the 1960's to characterize this competition more closely. They have detailed its advantages and disadvantages, and have established its connections with the ages and social status of the investigators, as well as with the scientific disciplines represented. Their analyses are, I believe, still very suggestive but problematic in their claims to generality, because they are ahistorical.

My intention in this paper is to show through an example that for historians it is relatively uninteresting to determine to which scientist each part of a

J. Meheus and T. Nickles (eds.), *Models of Discovery and Creativity*,
DOI 10.1007/978-90-481-3421-2_11, © Springer Science + Business Media B.V. 2009

discovery belongs, because discoveries are usually the result of collective transactions. I have chosen two examples from the field of vitamin research. They concern two vitamins, known today as B_2 and niacin. The case of vitamin B_2 involves dyestuff with the characteristics of an enzyme and a vitamin. The biologically active compound is called riboflavin—the most important derivatives of riboflavin are coenzymes of oxydases and dehyrogenases, FMN and FAD. Niacin is the coenzyme of NAD/NADPH. Both enzymes belong to the so-called hydrogen transport portion of the respiratory chain and are important for biological oxidations in the body cells. A lack of either vitamin causes severe deficiency symptoms in humane: lack of B_2 produces inflammatory changes of all sorts, lack of niacin produces pellagra.

Between 1932 and 1939 at least five scientists actively and bitterly disputed their respective shares in these discoveries. This time period extended from the first work of Warburg and Christian on the yellow enzyme (see Warburg and Christian, 1933) to the patent disputes over the synthesis of vitamin B_1. Several papers were associated with the discovery itself—they included the discovery of the biological effects of the substance, its preparation in crystalline form, its synthesis in the laboratory, and large-scale production. This consideration alone suffices to show the complexity of the application of the term "discovery" to vitamins. One more thing stands out: only in a few cases has the physiological mechanism of the effects been elucidated down to the fine structure of the cells. The two vitamins central to my discussion are, in this respect, exceptional.

In the historical sections of articles on vitamins in lexicons, handbooks of the history of science, or in current books about vitamins, evaluations of the role of individuals in the discoveries are very brief and stand in contradiction to the self-evaluations of the scientists. But their claims for themselves also underwent temporal changes. Thus the claims made during the controversies differed fundamentally from those made by the scientists in their later lives. One would expect the maturity of those involved, the passing of time, successes attained in the meantime, changing social contacts, and the further advance of knowledge to have an influence, but the effects were not those one would expect from such factors. I shall say more about that later.

2. The Significance of Collective Work

Hagstrom has examined with insight the various forms of competition. According to his definition,

> Competition results when two or more scientists or groups of scientists seek the same scarce priority-reward of discovery and the recognition awarded for it—when only one of them can obtain it. Competitors need not be aware of one another's existence. Collaboration occurs when two or more individuals consciously co-operate in seeking a scarce reward and share it, if and when it is obtained. (Hagstrom, 1965, p. 70)

He treats competition and collaboration as distinct, contrasting situations. In the example I present, however, phases of collaboration alternated with those of sharp competition, and the competitors were in contact with one another.

I would like first to introduce the scientists involved, and then ask the question whether the collaborative network was significant. The leaders of the scientific groups involved were each Nobel prize winners: Hans von Euler-Chelpin, Paul Karrer, Richard Kuhn, Hugo Theorell, and Otto Warburg. The groups of vitamin investigators up until about 1933 can be separated, both according to their methodologies and their directions of research, into two camps. Warburg (Berlin-Dahlem), his pupil and colleague Hugo Theorell (Stockholm), and the German Hans von Euler, residing in Sweden (Stockholm) concentrated on the elucidation of the functions of the coenzymes of the hydrogen transport enzymes of the respiratory chain which happened accidentally to be also colored. The clarification of the structure of the compounds whose functions they studied was, for these scientists, only of subordinate interest. In no case was the structure the starting point for their investigations.

The situation was different for Paul Karrer and his associates (Zürich), as well as for Richard Kuhn (Zürich, later Munich). As organic chemists, they concentrated their attention foremost on chemical structure, the synthesis of the compounds, etc.

Collaborative work ensued mainly on methodological grounds, because it was important to bring physical and chemical methods together to explain the structure and function of the compounds. Since 1927, for example, when Pohl reported for the first time the application of absorption spectra to the investigation of substances protective against rickets, spectroscopy had been ensured a prominent place in vitamin research, but only a few investigators applied the method. That was due, above all, to gaps in their training. Von Euler-Chelpin and Warburg, who had received strong training in physics and physical chemistry, used spectroscopy in their investigation of redox-systems. Warburg, on the other hand, seldom used chemical preparative method. Von Euler-Chelpin, in fact, strictly excluded from his laboratory the preparation of crystals, or the synthesis—in effect, the structural chemistry—of the vitamin derivatives he studied. He assured Karrer that he wished to stay out of Karrer's research field. Although it was expected initially that the detection and proof of the constitution of the vitamins, and their quantitative determination, would be possible by means of spectroscopy (in place of the time-consuming experiments on animals), the viewpoint gradually emerged that such methods could only supplement, not replace, the results gained by chemical methods (compare Rudy, 1936, p. 497). That was, above all, because one required reference substances that had to be synthesized. Moreover, in quantitative determinations one had to exclude by chemical methods the presence of other light-absorbing substances that were without vitamin action. Even in examining the constitution of the molecules, it was important to be cautious, because molecules that were chemically very different could display very similar absorption spectra. These insights provided the framework for an evolving collaboration between preparative and physical-chemically oriented chemists.

If one looks at the course of the collaborative work that began in 1933 and extended to about 1939, two phases can be distinguished, although they are hard to separate from one another. A first phase was, in my opinion, oriented around the acquisition of knowledge in the sense of basic research—the function and chemical description of lactoflavin and the vitamin today called niacin. In the second phase the synthesis of the substances became central. Even in the fist phase the collaboration did not go without conflict—there were frequent struggles over priority and breaking of norms. Through them a powerful dynamic emerged within the network, and emotional clashes ensued. Each scientist carried out his work with reference to that of the others, a dialogue took place that is comparable to a "resonance process"—something like the oscillations that arise when 100 soldiers march with identical footsteps across a bridge. In the second phase industrial considerations dominated. Industrial interests made collaboration more difficult, especially because the various investigators felt obligated to firms that competed with one another. This influence extended so far that it came to determine the choice of themes in the publications. The disputes ended in legal settlements.

3. How are the Results Evaluated from the Current Perspective?

I would like to turn first to the question, how are the contributions of the individual investigators evaluated from the perspective of today (see, among others, Bässler et al., 1992, p. 59). Here it is striking that the evaluations are very summary, and that groups that had competed with one another are named together without exercising more closely their respective parts (see Table 1). Thus the isolation and elucidation of riboflavin is broadly and summarily attributed to two working groups. To be sure, the authors of these judgments have been aware that there had been disputes. Thus, in an exposition of the work of Karrer it is stated that his "vitamin research took place in competition, and mutual successes, with that of Kuhn" (see Pötsch et al., 1989, p. 229). No mention is made of the competition over niacin, where there were also strong disputes. It is hard to find out why these events are so treated. Probably the cause lies in the little interest that historical events hold for natural scientists, but other kinds of explanation are also possible. The parts that individuals played in the advances have been forgotten, those concerned and their contemporaries have died.

The suspicion is, in fact, confirmed that contemporary referees placed greater value on the disclosure of the parts of individual scientists, and, therefore, can

Table 1. Modern evaluation of the contribution of individual researchers to the discovery of vitamines B_2 and niacin (Ammon and Dirscher, 1948, p. 182–183 = 1; Bässler et al., 1992, pp. 59–60 = 2; Pötsch et al., 1989 = 3)

Year	Vitamin	Contribution
1923		Euler and Myrbäck analyzed the enzyme of the fermentation (described by Harden and Young) and called it cozymase. When it was shown that the preparations that are called cozymase consist of 2 substances, von Euler called his enzyme codehydrase I—because it was known before Warburg's codehydrase. (1)
1932	B_2	Extraction of the yellow enzyme (FMN) by Warburg/Christian (2)
1933	B_2	Isolation of riboflavin by Kuhn/Weygand/Karrer (2)
1933–1934	B_2	Clarification of the structure and the synthesis by Kuhn/Weygand/Karrer (2)
1934	B_2	Theorell synthesized the yellow "Atmungsferment" purely and decomposed it into the coenzyme (FMN) and the apo-enzyme (protein) (3)
1935-1936	B_2	Partial synthesis of an enzyme out of lactoflavin, phosphoric acid and the protein part extracted from yeast by Kuhn. (3)
1936		Clarification of the cozymase by Euler and Warburg: the cozymase consists of 1 molecule nicotinic acid amide (Euler/Albers/Schlenk/Warburg/Christian), 1 molecule Adenin and 2 molecules pentose phosphoric acid (Euler/Schlenk) (1)
1938	B_2	Discovery of the FAD as a coenzyme of the d-amino acid oxydase by Kuhn (2)

be given historical precedence. Such attributions are given, for example, in the monographs of Ammon and of Dirschl. In spite of their detailed expositions, however, both authors were reticent in their evaluations of the case in dispute.

It is remarkable that with respect to nicotinamide the opinions of both authors differ from the self-evaluations of the concerned parties (see below). Ammon and Dirschl both dismiss Warburg's claim for priority in the identification of nicotinamide and present it as a collective achievement.

The self-evaluations of the scientists stand in contrast to these judgments (compare Table 2).

If we compare the two tables, the first thing that strikes us is that the concerned parties emphasized the differences in their contributions, and placed great value on establishing their respective parts exactly. In the case of niacin the different assessments of von Euler-Chelpin's role by his contemporaries is conspicuous. Each person estimated his own part to be greater than that of the others (see points 4–6). In the case of Karrer (see 4), other members of the network

Table 2. Contemporary evaluations of their parts in the discovery by the Nobel prize-winners themselves

Vitamin	Claim	Year	Source
1. B_2	Decomposition of yellow coenzyme (FMN and protein) and resynthesis by Theorell	1934	Warburg to Karrer on 27.11.1934
2. B_2	Synthesis: "On December the first, another note came out, sent by Karrer on October the 27th. It treats another subject, but contains the following undated 'remark with the correction': 'Lately someone succeeded in producing iso-alloxazine pigments that contain rests of sugar alcohols at the nitrogen atom at the 9 position.' It remains to be seen which claims of priority will be connected with this sentence."	1935	Kuhn, 1935
2a. B_2	Synthesis: Karrer claims the priority for the total synthesis		Karrer, 1950, p. 794
3. Niacin	Theorell adjudicated the priority of the synthesis of nicotinic acid amide to Warburg.	1935	Theorell to Warburg on 21.12.1935 (concerning niacin)
4. Niacin	Von Euler claimed the priority for the proof of Adenin and the nicotinic acid amide as well as the cozymase. Karrer: Euler is right, but Warburg has found nicotinic acid amide in the coferment earlier.	1936	Karrer to Warburg in May 1936 (concerning niacin)
5. Niacin	Warburg: Adenin in the coferment Euler: Pyridin in the Warburg coferment (including discovery of the substance and the impact equations)	1936	Warburg to Karrer on 11.5.1936 (concerning niacin)
6. Niacin	Euler's Cozymase: Warburg maintained that von Euler had only joined his concept (Warburg's concept) by analogism.	1948	Warburg, 1948, p. 24
7. Niacin	Karrer adjudicated the codehydrase II (= TPN) to Warburg and the cozymase (= codehydrase I = DNP) to von Euler.	1950	Karrer, 1950, p. 796

evaluated Karrer's role generously. The matter concerned, ultimately, a quarrel between Von Euler-Chelpin and Warburg, with which Karrer had nothing to do. This situation leads one to suspect that social connections played such a role in the evaluations that objectivity did not come much into play. A statement to Warburg by Karrer in 1936 supports this view:

> As you know, I have been connected through friendly relations with Prof. v. Euler through many years of collaborative work, and I would not like to expose this connection to injury, least of all because of a scientific question. I believe, therefore, that I can continue to work together on the co-enzyme problem only if a friendly resolution can be found for the differences that have arisen between you and the Stockholm laboratory. (letter of Karrer to Warburg, undated, probably May 1936, NL Karrer, Switzerland)

The expectation that the wisdom brought by the passage of time, or a broader overview of the result, could lead to a more differentiated and tolerant view, was not realized in the case of Otto Warburg. Whereas he had earlier allowed his competitor a part in the discovery (compare 5), he later maintained that von Euler-Chelpin had obtained his results solely by drawing conclusions by analogy from Warburg's own results. This harshness fits with Warburg's reactions in other cases that cannot be treated here, for example in his quarrels with David Keilin.

Kuhn was, on the other hand, later judged more mildly by Warburg, because of the better relations that developed afterward between them. He even admitted that with the identification of lactoflavin—or riboflavin, as Kuhn named the substance—the prosthetic group of the yellow enzyme seemed to have been isolated. Warburg emphasized, however, that this proved to be an error, and that Theorell was the first to find out that the substance was actually riboflavin phosphate. According to Warburg, Kuhn confirmed this finding in 1936, when he repeated Theorell's synthesis experiments and produced the yellow enzyme with synthetic acid.

A separate chapter in the contests between these scientists were the so-called materials and procedures claims. Scientific priority claims were then joined to financial interests. Here Karrer established his claim over Kuhn for the total synthesis of B_2 as late as 1950.

These conflicts began at the end of the thirties. At first Warburg and Karrer stood together, against Kuhn, but later only Karrer and Kuhn opposed one another. Their differences broke out over a decomposition product which all sides suspected to contain the active vitamin. The matter involved first the priority for the identification, and second for the synthesis. In 1933 Warburg and Christian had discovered that in alkaline solution flavin showed a "photolytic" reaction, in which the action of light produced a substance soluble in chloroform that they named "lumiflavin". They suspected that the activity of flavin arose from a transformation carried out in the cells. Finding the active substance would have meant, besides the identification of the vitamin, the discovery of a fundamental life process. Kuhn also occupied himself with this problem, and was able to show that in the transformation a carbohydrate residue was split off (compare Warburg and Christian, 1933, pp. 228–229). By illuminating lactoflavin in the presence of air in neutral or acid solution, Karrer found a bluish florescent derivative that he called "lumichrome". A controversy arose over these matters between Kuhn and Karrer. Karrer rejected Kuhn's claim to have published the first knowledge about the structure of the new product of illumination, and stressed that he and his colleagues had submitted the communication "Lumichrome, a new Product of the Illumination of Lactoflavin", to *Helvetica Chimica Acta*, on July 11, 1934. At this time nothing certain was yet known about the chemical structure of the compound. Kuhn and his co-workers had suspected, 4 to 8 weeks earlier, that it was a methylamide compound. While Karrer's paper was in press, a new publication from Kuhn and his associates appeared, which repudiated the view they had expressed a few weeks earlier and suddenly announced the synthesis of lumiflavin. Karrer took this affair so seriously that he expressed his view in 1934 in a special publication in the "Berichten der deutschen chemischen Gesellschaft" (see Karrer, 1934). He and his co-workers had already produced a flavine of the same type, with side-chains containing hydroxyl groups, and had investigated its properties in light in neutral and alkaline solution, before the publication of Kuhn appeared.

> When we were able to show in [...] our paper that lactoflavin is decomposed by light to 6,7-dimethyl-alloxan (lumichrome), the first secure foundation for the formula of lactoflavin was given. (Karrer, 1934, p. 2061)

Karrer pointed out that his work had been submitted earlier, and defended himself from Kuhn's claim that lumichrome had already been discovered earlier by him in the form of an impure preparation, as the product derived from the illumination of lactoflavin (see Karrer, 1934). Warburg commented at the time on this dispute, with reference to Kuhn:

> Even if one invokes the struggle for existence as a mitigating circumstance, one should not permit the further spread of such methods. (Letter of Otto Warburg to Paul Karrer, January 10, 1935; private archive of Heinz Karrer, NL Paul Karrer, n.f.)

The "methods" Warburg had in mind was the fact that Kuhn obviously had received from the work of a colleague the suggestion on which he based his own investigation. Karrer disapproved of this behavior as much as Warburg did. He regarded it as intruding in their field of research. Here too, however, there were divergent opinions. Kuhn's co-worker Wagner-Jauregg presented the situation as though von Euler-Chelpin, together with Karrer, had intruded in Kuhn's research field. Together with von Euler-Chelpin, Karrer had allegedly begun the isolation of vitamin B_2 independently of, but later than Kuhn. Wagner-Jauregg recalled that:

> When he [Kuhn] saw the yellow-green florescent solutions standing on my laboratory bench, he was visibly startled. Karrer was certainly present at the lecture on the discovery of riboflavin that I presented at the meeting of the Schweizer Naturforschenden Gesellschaft in Altdorf. (Wagner-Jauregg, 1985, p. 45)

However that may be, the right to establish claims must be conceded to all sides, all the more because each of them made important contributions.

This quarrel between Kuhn, Warburg, and Karrer proved retrospectively to have been extremely productive for the further development of the science. These and other controversies within their narrow field led to conversations about setting up boundaries between the respective research areas. There were, for example, suggestions that one person leave to another the further investigation of a particular substance (for example garderin, lactoflavin) or research problem (for example Szent-Györgyi). These attempts failed, all the more because those who sought to establish such norms in one case had not held to them in other cases, or did not stick to them. All of those involved recognized—correctly as it turned out—that they were dealing with a very important compound. Now it came to protecting specific synthetic steps. It turned out that Kuhn proved himself to be, in this respect, very clever, in that he actually patented each step. This strategy later played a role, in 1939, in the patent negotiations between I. G. Farben and Hoffmann-La Roche.

On December 9, 1938, the firm Hoffmann-La Roche, with whom Karrer collaborated, was denied a German patent application for the production of lactoflavin. A patent application of Hoffmann-La Roche for the preparation of intermediary products of lactoflavin synthesis, the so-called isoalloxazine derivatives, had a similarly contradictory outcome (granted in Belgium, Denmark, Poland, Sweden, Switzerland, and Great Britain; denied in Germany, Holland; withdrawn in Austria).

It is notable that in the negotiations between the firms all the vitamins—not vitamin B_2 alone—were treated as a "package". National and foreign patents, as well as patent applications, were evaluated against one another. On February 15, 1939, a settlement was reached between representatives of I. G. Farben and Hofmann-La Roche. As can be inferred from rough drafts attached to the protocol, the representatives of the firms confronted one another like hostile powers. After an exchange of aggressive attacks (a letter from the general director of Hofmann-La Roche, Emil Barell was characterized as "rude Tobacco", an intolerable Swiss cigar) they cut the Gordian knot and forged a compromise. This was possible because the firms were protected differently in the patent field, and their central need was to reach a balance of their respective interests. For example, they balanced vitamin E (then called aneurin) against B_2 (then lactoflavin). After both sides had fought a tough poker game with all the means at their disposal (for example one concealed the inadequacy of ones own patent rights by counting up foreign patents, or did not answer certain questions), they reached a balance of interests in 1939. From the side of Hoffmann-La Roche the following suggestion was made:

> Both partners offer each other reciprocal free license on their present and future patent rights in the lactoflavin field. Roche commits itself to an offer of license sharing with I. G. in the vitamin E field, whereby the royalty paid to Roche would be determined according to the rights possessed by the two sides.

Live and let live was the clearly stated message. In this "package deal" the question that no one could rightly answer was always at the center of concern—that is, of the importance and commercial significance of the vitamins. Which ones were irrelevant, and which ones were not?

Table 3. Conflicts that ended in patent disputes

Kuhn/I. G	**Claim for substance of I. G. Farben for intermediate of the synthesis of Lactoflavin (= 1-Ribityl-amino-2-amino-4,5-dimethylbenzol)** Publication of Kuhn in the "Berichten der deutschen Chemischen Gesellschaft" in november 1934 (concerning the arabityl derivate and other derivatives of the Diamino-xylids, but not its ribityl derivate). Reference: Publication of Kuhn in November 1934 about arabityl derivates and derivatives of the Diamino-xylids.	1938/39	Lactoflavin/B_2
Karrer/La Roche	**Claim for substance 1-Ribityl-amino-2-amino- 4,-dimethylbenzol** Reference: Publication of Karrer in February/March 1935	1938/39	Lactoflavin/B_2

The question is of interest, why exactly in the field of vitamin research did such bitter competition arise? This dispute is, I believe, understandable only if one looks at it as a controversy over norms. The system of scientific norms established by Merton (1985) can be criticized in the same way that many authors have criticized communism and organized skepticism (which attempts to prevent the dogmatisation of knowledge)—because of its claim to timeless validity. Scientific conclusions can only be established when they are confirmed by the scientific community. One of the norms in question was respect for the research areas of others. Breaking such rules led to the conflicts described in this network. But in the 1950's, for example, there were strong national differences in the observance of this norm. James Watson recalled in 1968 in his book "The Double Helix", that at the end of the forties and beginning of the fifties in Great Britain, the rules of "fair play" inhibited him and his colleagues in Cambridge from turning to the molecular investigation of DNA, because this was the theme of the research of their colleagues at King's College in London. Watson attributed this situation to norms prevailing in England that were unknown in France or the United States. This became especially clear in comparison with the activities of the famous Linus Pauling, at Caltech, for whom such rules did not hold. With competition becoming more intense in all fields, among other reasons because of ties to industry, the protection of research areas no longer played a role. With respect to keeping individual results secret there have been and are differences between nations, and in the case of the United States even between the East and the West coast. Thus Hagstrom found that many of those involved in research in Europe felt a greater need to conceal results than did those in the USA, and again those on the East coast of the USA more than those on the West coast (Hagstrom, 1965, p. 89).

The community of vitamin researchers was, on the other hand, not only international, but also interdisciplinary (including organic and physical chemists, cell physiologists, and physicians). Consequently many scientists pressed into the field of vitamin research, the key questions were clearly defined with regard to the explanation of the structure and functions of the vitamins, and there was a common definition of the significance for these questions of their synthesis or isolation, so that they all came into direct competition in a narrow field. New arrivals were not easily accepted by the established investigators. Here hierarchies, as well as the cycles of recognition for past achievements and credibility described by Latour and Woolgar, also played a role.

How do I, as an outsider, evaluate the respective contributions of each investigator in this case? That is very difficult. There are many problems. First, one has to take into account subsequent changes in concepts and notation. For example, NADPH/NADP was earlier called TPN or DPN. But there are more fundamental problems in identifying criteria for such an evaluation.

Should the idea, the theory, or the experiment count? Is there an overall "idea"? Does the story of discovery not present rather a complicated sequence of necessary errors? To say nothing about so complicated a theme as "What is a proof"? Is a graphical presentation accepted as such a proof? Or a crystalline substance? A system of proofs free from contradictions? And when one has decided to recognize an intermediate step as essential, what formal criterion for priority should one follow—the date of submission of a publication? The date of its acceptance? The date of appearance of the publication? Regarding even such formal criteria, I encounter the difficulties already mentioned, in the example of the dispute between Karrer and Kuhn, where there are discrepancies between the time of acceptance and of publication. Papers that were accepted later sometimes appeared earlier.

The role of ideas seems to me more interesting. I would like to recall what Sir Lawrence Bragg wrote in 1969, using the example of the investigation of DNA, about the mutual interactions of collaborators (see Watson, 1969, pp. viii–ix).

> It is not easy to be sure whether the crucial new idea is really one's own or has been unconsciously assimilated in talks with others. The realization of this difficulty has led to a somewhat vague code amongst scientists which recognizes a claim in a line of research staked out by a colleague—up to a certain point. When competition comes from more than one quarter, there is no need to hold back.

In a memoir that appeared in 1985, Kuhn's colleague Wagner-Jauregg expressed uncertainty over from whom the ideas about working on vitamin B_2 originated—whether from Kuhn or from Paul Györgyi (see Wagner-Jauregg, 1985, p. 42). Györgyi, on the other hand, as a letter from him to von Euler-Chelpin shows, was certain that the investigations were based on his ideas and took place through his initiative.

It is, accordingly, difficult to assess contributions and priorities. In the case of niacin it seems to me that Warburg was right in his claims about von Euler-Chelpin. Warburg's argument was that von Euler-Chelpin had changed his view under the influence of Warburg's work, and came suddenly, after ten years, to believe that cozymase was a dinucleotide. His shift can, therefore, be traced back to a stimulus from Warburg. On the other hand, von Euler-Chelpin had also carried out his own experiments, although they had not yielded an empirical formula conforming exactly to this composition. The accurate formula was, in fact, proven in 1936 by Warburg and his associates. In the case of vitamin B_2 the situation is less ambiguous. Theorell found the proethetic group. Kuhn repeated the experiment. The claim by a co-worker of Karrer that Kuhn was the first to succeed with the synthesis is false.

The patent struggles were only indirectly connected with the priority struggles. The problem was that Kuhn was able to obtain patent protection for certain intermediate products that Karrer had also used in his synthesis.

4. How Effective was the Network?

The task of evaluating the effectiveness of the network seems to me to be easier. The evaluation can be concerned only with the results. The vitamin B_2 and the niacin that were contested are those whose physiological function are today well known. They are enzymes with the properties of dyestuffs and of vitamins. That means that the conjunction of the work of the groups that concerned themselves primarily with physiological function with the work of those who concentrated on structure produced an achievement that is still recognized. That results were published relatively rapidly can be attributed to the intensity of the competition. Kuhn, for example, published more than 70 articles on vitamin B_2. The pressure to publish intermediate results led to the rapid spread of information about the state of the work. Duplication of work was, in this way, avoided. Each person sought to develop the most efficient means of synthesis and to find ways around key substances used by competitors.

It is not easy to answer the question whether the reorientation of scientific interest around the synthesis was productive or not. On the positive side, in any case, is that, in the end, it was due to the industrial patents that vitamins could be prepared industrially relatively early and made available to the people.

5. Conclusion

I have shown that the individual contributions of investigators to important discoveries have been sources of dispute and misunderstanding. The assessments of the participants differed sensibly from one another. Here social factors such as friendships, enmities, and previous disputes played a role. Later evaluations were mostly summary.

I would like to return to the position I took at the beginning, that it is relatively meaningless to try to establish the individual contributions to a discovery, because individual achievement can only be evaluated as part of the interactions of a group. Concerning the nature of discovery, Thomas Kuhn has written

> Examining selected discoveries, we shall quickly find that they are not isolated events but extended episodes with a regularly recurrent structure. Discovery commences with the awareness of anomaly, i.e., with the recognition that nature has somehow violated the paradigm-induced expectations that govern normal science. (Kuhn, 1970, pp. 52–53)

Leaving aside the particular association that Kuhn makes between discovery and the framework of paradigm expectations that govern his view of normal science, his assertion that discoveries are events extended in time impinges also on the question of their attribution to individual scientists. He asks, for example, when oxygen was discovered and whether it was Priestley or Lavoisier who discovered it. He writes, "any attempt to date the discovery must inevitably be arbitrary because discovering a new sort of phenomenon is necessarily a complex event, one which involves recognizing both that something is and what it is" (Kuhn, 1970, p. 55). These problems, already evident in defining a discovery of the late eighteenth century, are far more perplexing when we apply our analyses to discoveries such as those I have described, made during the intensely competitive conditions that prevail in the twentieth century. Our efforts to elucidate the processes of discovery in science must take into account the prior difficulties involved in identifying when a discovery has taken place and who has taken part in it.

References

Ammon, R. and Dirscher, W. (1948). *Fermente, Hormone, Vitamine und die Beziehungen dieser Wirkstoffe zueinander*. Georg Thieme, Leipzig. Second edition.

Bässler, K.-H., Grühn, E., Loew, D., and Pietrzik, K. (1992). *Vitamin-Lexikon*. Jena, New York, Stutgart.

Hagstrom, W. O. (1965). *The Scientific Community*. Basic Books, New York/London.

Karrer, P. (1934). Bemerkungen zu Abhandlungen von R. Kuhn und Mitarbeitern über Flavine. *Berichte der deutschen chemischen Gesellschaft*, 67:2061– 2063.

Karrer, P. (1950). *Lehrbuch der organischen Chemie*. Georg Thieme, Leipzig. Tenth edition.

Kuhn, R. (1935). Bemerkungen zu Abhandlungen von Paul Karrer und Mitarbeitern über Flavin. *Berichte der deutschen chemischen Gesellschaft*, 68:173– 176.

Kuhn, T. (1970). *The Structure of Scientific Revolutions*. University of Chicago Press, Chicago. Second, enlarged edition.

Latour, B. and Woolgar, S. (1979). *Laboratory Life. The social Construction of Scientific Facts*. Sage, Beverly Hills.

Merton, R. (1985). *Entwicklung und Wandlung von Forschungsinteressen*. Aufsatze zur Wissenschaftssoziologie, Frankfurt/Main.

Pohl, R. (1927). Zum optischen Nachweis eines Vitamines. *Die Naturwissenschaften*, 15:433–438.

Pötsch, W., Fischer, A., Müller, W., and Cassebaum, H. (1989). *Lexikon bedeutender Chemiker*. Leipzig, Frankfurt am Main.

Rudy, H. (1936). Absorptionsspektren im Dienste der Vitaminforschung. *Die Naturwissenschaften*, 24:497–505.

Wagner-Jauregg, T. (1985). *Mein Lebensweg als bioorganischer Chemiker*. Wissenschaftliche Verlagsgesellschaft, Stuttgart.

Warburg, O. (1948). *Wasserstoffübertragende Fermente*. Saenger, Berlin.

Warburg, O. and Christian, W. (1932). Über ein neues Oxydationsferment und sein Absorptionsspektrum. *Biochemische Zeitschrift*, 254:438–458.

Warburg, O. and Christian, W. (1933). Über das gelbe Oxydationsferment. *Biochemische Zeitschrift*, 263:228–229.

Watson, J. D. (1969). *Die Doppelhelix*. Rowohlt, Hamburg.

Werner, P. (1998). *Vitamine als Mythos*. Akademie Verlag, Berlin.

Index

"a-ha" experience, 52
abduction
    -deduction-induction cycle, 100
    and affirming the consequent, 100
    and belief revision, 106
    and deduction, 99, 104
    and discovery, 82
    and inconsistencies, 96, 104–114
    and induction, 99
    and inference to the best explanation, 97, 99, 100
    and prior appraisal, 82, 83
    and problem solving, 82, 99
    and theory change, 100
    deductive models of, 104–106, 110, 111, 113
    heuristic constraints on, 83–86
    in artificial intelligence, 99
    kinds of
        creative, 97–99, 101, 109–114
        manipulative, 103, 104, 104*n*
        model-based, 99, 103, 111
        selective, 97–100
        theoretical, 97–99
        visual, 101–103, 111, 112
abductive arguments, *see* arguments, abductive
abductive reasoning, *see* reasoning, abductive
acceptance
    and graded structures, 4
    of changes in taxonomies, 4, 23, 24
    of new discoveries, 3, 23
    of the nuclear fission hypothesis, 2, 22, 23
action-at-a-distance, 149
ad hoc hypotheses, 122
adaptation
    biological, 138, 182*n*
    socio-cultural, 136, 138
affordances, 133, 134, 136, 155, 157
Agruss, M., 1, 12, 12*n*, 14–16, 22, 24
AI, 102, 108, 128, 129, 132, 172, 175, 176, 176*n*, 178, 190, 193, 198, 213
Alchourrón, C., 104
algebra
    introduction of complex numbers in, 58, 60–62
    thought-experiments in, 60–62
Altenberg, L., 186
Amaldi, E., 6, 24
Ammon, R., 227
Ampère, A. M., 131
ampliative logics, 183*n*
analogical modeling, *see* modeling, analogical
analogical reasoning, *see* reasoning, analogical
analogies
    abstract, 85, 86
    and causality, 86
    and conceptual change, 148, 152
    and explanatory promise, 85
    and generic abstraction, 152
    and mechanisms, 91
    and problem solving, 102–103
    and relation structures, 154
    and the electromagnetic field concept, 85, 147, 149, 160
    and the introduction of new concepts, 35
    and the introduction of new entities, 35
    and theory construction, 91
    causal, 86
    for natural selection, 85
    for the nucleus, *see* droplet analogy
    material, 85, 86, 90, 91
    Maxwell's physical, 85
    physical, 154
analytic-synthetic distinction, 34
Andersen, H., 2, 3*n*
Anderson, A., 97
anomalies
    and consistency maintenance, 113
    and constraints, 45, 50
    and mechanisms, 44, 50
    and taxonomies, 3, 4, 23, 24
    and the discovery of nuclear fission, 4, 22–25
    and the discovery of the mechanism for protein synthesis, 51–53
    and the explanation of dinosaur extinction, 84

and theory change, 111
conceptual, 111–112
diverging assessments of, 3, 23–25
empirical, 81, 111–112
in mechanism discovery
compositional, 50, 51
temporal, 50, 51
in the periodic table, 38
localization of, 50, 51, 54
resolution of, 44, 50–54, 111, 112
severeness of, 3–4, 23, 25
triggering new discoveries, 3
anomaly-driven revision, 44
anthropology, cognitive, 128
Apostel, L., 216
appraisal, prior, 82, 83
Archimedes, 57, 58, 60, 63, 64
Archimedes's method, 58–60
argumentation
abductive, 92
in mathematics, 58
arguments
'after the facts', 58, 59, 62, 63
abductive
conditions for, 86
for natural selection, 85
and thought-experiments, 63, 152
by analogy, 152
deductive, 138, 188
demonstrative, 63
discovery, 176*n*
in mathematics, 58
inductive, 138
Aristotelian physics
concepts of earth, water, air, 29, 33, 37–38
concepts of fall and natural place, 36*n*
concepts of motion, 37–38
Aristotle, 167, 169, 186
artificial intelligence, *see* AI
Astrachan, L., 52
automated discovery, 195
automatically defined functions, 195

Bässler, K.-H., 226
Bachelard, G., 217
backward chaining, 175
BACON, 175*n*
Bacon, F., 168–173, 176, 180, 200, 203
Baconian methods, 173–174
Balzer, W., 219
Barsalou, L. W., 4*n*, 143, 157
Batens, D., 109, 183*n*
Bayesianism, 112
Becher, V., 104, 106, 108
Bechtel, W., 45
belief change
and conceptual change, 109
and inconsistencies, 108
and suppositional reasoning, 104*n*
coherence approach to, 108, 109
foundations approach to, 108, 109
kinds of, 104, 106
belief revision
and abduction, 106
formal framework for, 106–109
probabilistic, 112
Belnap, N., 97
Belozerskii, A. N., 51
Bernard, C., 70–72, 77
biology
developmental, 86
discovery in, 45
evolutionary, 82, 86–92
molecular, 44–46, 49, 50, 92
black box incompleteness, *see* failures (in mechanisms), incompleteness
Black's law, 175*n*
Black, J., 176*n*
Black, J. B., 144, 145
blind variation, *see* BV+SR
Bloom, H., 187*n*
Bloor, D., 177*n*
Bobrow, D. G., 144
Boden, M., 128, 209, 212, 214
Bohr, N., 20, 22
Bombelli, R., 61, 62, 64
Borges, J. L., 191, 191*n*, 201*n*
Boutilier, C., 104, 106, 108
Boyle, R., 67
Bragg, L., 234
Brandon, R., 45
Brannigan, A., 175*n*
Brenner, S., 52, 53
Brewka, G., 107
British museum algorithm, 190, 191
Britten, R. J., 88
Brown, H., 29*n*, 34*n*, 36*n*
Brown, J. S., 132, 140
Bruzzaniti, G., 38
Burian, R. M., 45, 46, 51, 54
Butterfield, H., 169*n*
BV+SR
algorithmic characterization of, 192
and biological evolution, 192, 197
and Campbell, 189
and chance, 185
and constraints, 186*n*
and creativity, 187, 193
and innovation, 176, 183*n*, 185, 186
and learning, 176, 186, 197*n*
and method, 190, 193
and search, 189
and the growth of knowledge, 199
model of inquiry, 179, 182, 189, 201–202

Bylander, T., 104
Byrne, R., 144, 189

Campbell, D. T., 176–179, 183*n*, 184–186, 188–190, 197*n*, 199, 201
Cantor, G. N., 159
Cartesian methods, 173–174
Cartwright, N., 153
case-based reasoning, *see* reasoning, case-based
causality
and analogies, 86
and heuristic criteria, 84–85
Chamberlain, T. C., 202
chance, *see also* luck
and discovery, 168
and innovation, 172, 178
and inquiry, 172
and method, 168, 178, 180
chemical elements, *see also* transuranic elements
missing in the periodic table, 23
taxonomy of, 5, 6, 19
chemistry
introduction of isotopes in, 29, 38
nineteenth-century, 38, 39
Chen, W., 74
Chen, X., 3*n*
Chi, M. T. H., 140
Christian, W., 224, 229, 230
Church, A., 117
Clark, K. L., 115–117
classical discovery program, *see* discovery program, classical
Clement, J., 140
cognition
and action, 133
and culture, 131–137, 158–162
and environment, 135
and external representations, 136
and mental modeling, 140
and socio-cultural factors, 130–131
contextualized accounts of, 132
distributed, 131–137, 139
interactionist approach to, 136
logicist accounts of, 132
naturalistic account of, 177
non-reductionist analyses of, 131
situated, 131–137, 139
social accounts of, 130
traditional account of, 129–130, 132
cognitive-historical analysis, 127–131, 137, 138, 146
coherence, *see also* heuristic criteria, coherence
connectionist models of, 113
explanatory, 96*n*, 112, 114
of conceptual systems, 113
collaboration, *see* discovery, and collaboration
Collingwood, R. G., 217
competition, *see* discovery, and competition
computation
evolutionary, 167*n*, 176, 190, 193, 193*n*, 194, 197, 198, 202, 203
knowledge-based, 176
computational philosophy, 102
concept formation, 102, 103, 127, 159
concept learning, *see* concepts, learning of
concepts, *see also* graded structures
analogous, 35
and actions in the world, 34*n*
and conceptual roles, 36
and conceptual systems, 29, 31
and departure transitions, 31, 36
and entry transitions, 30, 31, 33, 33*n*, 35, 36, 38, 40
and habits, 31–33, 36
and non-linguistic beings, 31
constraints on a theory of, 34
content of, 29, 34
empiricist accounts of, 30
holistic accounts of, 29
introduction of new, *see* conceptual innovation
Kuhn's theory of taxonomic, 3
learning of, 30, 32, 34*n*
naturalistic accounts of, 32
necessary-and-sufficient-conditions view of, 34*n*
non-analytic, 34*n*
of earth, water, and air
in Aristotelian physics, 29, 33, 37–38
in Galileo's dynamical theory, 29, 36–38
of fall and natural place
in Aristotelian physics, 36*n*
in Galileo's dynamical theory, 36–37
of logic and mathematics, 30
of motion
in Aristotelian physics, 37–38
in Galileo's dynamical theory, 36–38
of space and time, 30
open-textured view of, 34*n*
paradigm instances of, 30
representations of, 138, 140
Sellars' theory of, 29, 35, 36
Sellarsian approach to, 29, 36, 38, 40
similarity account of, 3, 25
types of
descriptive, 30, 31, 33, 34, 34*n*, 35, 36
formal, 30
normative, 31
conceptual change
and analogies, 152
and belief change, 109
and continuity, 29
and creativity, 112, 127, 137–153

and inconsistencies, 110
and model-based reasoning, 137–153
and models, 138
and problem solving, 146
and thought-experiments, 151, 153
and visual modeling, 149
H-creative, 128
linguistic perspective on, 138
P-creative, 128
Sellarsian approach to, 32–35, 40
conceptual comparison
and the content of concepts, 29
between Aristotelian physics and Galileo's dynamical theory, 29, 36–38
conceptual innovation
and analogical modeling, 148
and creativity, 127
and flashes of insight, 146
and generic abstraction, 149
and model-based reasoning, 138, 139, 153
and reasoning, 146
and the content of concepts, 29
as a problem-solving process, 138
cognitive-historical investigations of, 129, 138
in chemistry, 29
in Maxwell's thinking, 153
Sellarsian approach to, 34, 35, 40
conceptual practices, 127–131
conceptual problems, 111–112
conceptual structures
and creativity, 209
and models, 138
individual differences in, 2, 3
representations of, 138
taxonomic, *see* taxonomies
conceptual systems
and habitual inferences, 32
and implications, 31, 33
and language, 31
and their extra-systemic subject matter, 31
coherence of, 113
of space and time, 30
types of, in Sellars' approach, 30
conjectures and refutations, 177, 190*n*
connectionism, 176
consensus
and the similarity account of concepts, 25
Kuhn on, 81, 211
on the circulation of the blood, 75
consequentialism, 174
consilience, *see* heuristic criteria, consilience
consistency, *see* heuristic criteria, consistency
consistency-based reasoning, *see* reasoning, consistency-based
constraints
and affordances, 133, 134, 136, 155, 157
and anomalies, 45, 50, 54
and anomaly resolution, 52
and unexpected discoveries, 2
attunement to, 133
guiding hypothesis evaluation, 45
guiding hypothesis generation, 44–45
guiding mechanism discovery, 45
in the discovery of messenger RNA, 53
individual differences in, 23
non-identical, 2, 25
on the organization of mechanisms
componency, 45, 46
compositional, 45
hierarchical, 46
spatial, 46
temporal, 45, 46, 53
violation of, 2, 25, 45, 50
contradictions, *see* inconsistencies
controversies, scientific, 96*n*, 224, 230–232
conventionalism, 114, 117, 119, 122
conventions
and *ad hoc* hypotheses, 122
geometrical, 118, 118*n*
withdrawal of, 114, 117–122
Cooper, L. A., 142, 144
Copernicanism, 37
Craik, K., 140–142, 144
Craver, C., 43, 45, 47, 49, 50, 52, 54
creative processes, fine structure of, 69
creativity
and BV+SR, 193
and conceptual change, 127, 137–153
and conceptual innovation, 127
and conceptual structures, 209
and genius, 127, 187
and innovation, 210–211
and inspiration, 180
and method, 180
and planning, 187
and representational change, 146
and socio-cultural context, 158
and the semantic view of theories, 212–214
BV+SR model of, 187
evolutionary model of, 184, 187
explore-and-transform-paradigm of, 209, 210, 212, 216–218
God design model of, 185, 187, 189
H-, 128–129, 212
human design model of, 183, 186–189, 192*n*
in the applied sciences, 214–216
in the mature sciences, 216–218
instructionist theories of, 187
P-, 128–129, 212
providential theories of, 187
romantic models of, 187

Crevier, D., 176*n*
Crick, F., 49, 51–53
Crookes, W., 38
crucial experiments, 65
Curie, I., 4
Curie, M., 120
Cziko, G., 185, 186*n*, 189, 192*n*

D'Agostino, O., 6
Dalton, J., 38
Darden, L., 43–45, 47, 49, 50, 52, 85, 111, 129, 138
Darwin, C., 85, 127, 177, 181, 183–186, 190, 197, 198
Davidson, E. H., 88
Dawkins, R., 177, 182, 186*n*, 189
De Beer, G. R., 89
de Kleer, J., 104, 106, 108, 140
deduction
  and abduction, 99, 104
  and diagnostic reasoning, 100
  and induction, 99
Delbrück, M., 185
demarcation, 169
Dennett, D., 177, 186*n*, 187, 189, 191, 192*n*, 197*n*, 199
Derrida, J., 68
Descartes, R., 62, 168–176, 180, 181, 187, 200, 203
diagnosis
  consistency-based, 107, 108
  medical, 97, 100
  model-based, 106
diagnostic reasoning, *see* reasoning, diagnostic
diagrams
  and mechanisms, 44, 46, 50, 51
  and model-based reasoning, 103, 155
  and problem solving, 135
Didion, D., 54
Dietrich, M. R., 82
Dijksterhuis, E., 59
dinosaur extinction, 85
Dirscher, W., 227
discoverability, 173*n*, 176*n*, 184*n*
discoveries
  acceptance of new, 3, 23
  and unprecedented events, 73, 74, 76
  as social constructions, 175
  different claims to new, 4
  expected, 12
  explaining particular, 169, 170
  multiple, 220, 223
  neglect of new, 3
  revolutionary, 23
  unexpected, 2
discovery
  and abduction, 82
  and anomalies, 3
  and chance, 168
  and collaboration, 224–226
  and collective work, 223–226
  and competition, 224–226
  and constraints, 53
  and heuristic pluralism, 82
  and hypothesis generation, 52
  and innovation, 175
  and justification, 44, 63, 171, 174, 184
  and luck, 168, 170*n*
  and mechanisms, 43–46, 53
  and problem solving, 43, 53, 175
  and reasoning strategies, 53
  and scientific method, 169
  and thought-experiments, 60, 63
  and visual thinking, 112
  as a process, 44, 53
  attribution of, 175*n*, 235
  automated, 195
  context of, 44, 184*n*, 190, 190*n*, 213
  evolutionary model of, 184, 187
  friends of, 193
  history of, 168
  human design model of, 183, 186–189, 192*n*
  in biology, 45, 53
  instructionist theories of, 187
  logic of, 46, 54, 102, 171, 173, 177
  machine, 203
  method of, 168, 171, 175*n*, 176, 177, 181, 184, 198, 201, 203
  methodoloy of, 172
  methods, 95, 102
  of complex numbers, 58, 60–62
  of gene regulation, 88
  of Krebs cycle, 72
  of mechanisms, 45, 46, 49, 53, 54
  of messenger RNA, 52
  of multiple decay processes, 20
  of niacin, 224
  of non-Euclidean geometries, 112
  of nuclear fission, 2, 2*n*, 4, 22
  of oxygen, 175*n*, 235
  of penicillin, 74
  of quantum theory, 175*n*
  of radioactivity, 39
  of RNA, 74
  of the area of the segment of a parabola, 58–60
  of the circulation of the blood, 74, 75
  of the citric acid cycle, 77
  of the Euler formulas, 62
  of the mechanism of protein synthesis, 49, 52
  of the neutron, 4
  of the ornithine cycle of urea synthesis, 72

of the structure of DNA, 49, 50
of the theory of the gene, 43
of theories, 44, 46
of transuranic elements, 12, 16, 23
of vitamin B_2, 224
providential theories of, 187
romantic models of, 187
socio-cultural context of, 217
traditional views of, 169–171
discovery program, classical, 171, 173, 178, 179, 181, 200–204
discovery, mathematical
and thought-experiments, 58
logic of, 58
disintegration processes, taxonomy of, 5–8, 12, 13, 16, 19, 20, 22–24
dissensus
and the similarity account of concepts, 25
Kuhn on, 81
distributed reasoning, *see* reasoning, distributed
DNA replication, mechanism of, 46
DNA synthesis, mechanism of, 49
DNA, discovery of the structure of, 49, 50
Donald, M., 136
Dostoevsky, 191
Doyle, J., 108, 109
droplet analogy
Bohr's, 20, 22
Gamow's, 16
Duhem, P., 168*n*, 217

Edelman, G., 185, 186*n*, 189
Einstein, A., 127, 177, 188
Eldredge, N., 87
electromagnetic field
concepts of the, 102–103, 150, 153, 158–160
laws of the, 154
embodiment, 152
epistemology
evolutionary, 172, 176–178
foundational, 169, 171
Euclid, 60
Euler, 62
evaluation, *see* reasoning strategies, for evaluation
evolution
and innovation, 182
biological, 181–182, 184
cognitive, 136
cultural, 135–137
evolutionary computation, *see* computation, evolutionary
evolutionary epistemology, *see* epistemology, evolutionary
evolutionary model, 184, 187
experiment
and theory, 65–67
crucial, 65
Michelson-Morley, 65
Millikan oil drop, 65
experimental
science, 67, 72
systems, 68–69, 71–73, 78
experimentation, 65–67, 73
explanation
covering law model of, 109
statistical, 111
explanatory power, *see* heuristic criteria, explanatory power
explanatory promise, 82

failures (in mechanisms)
fixing, 47, 49
incompleteness, 43–45, 47, 53
incorrectness, 43, 44, 50
kinds of, 47
localization of, 44, 50
Fajans, K., 39
fallibilism, 171, 173, 184*n*
Faraday, M., 102, 131, 150, 159
Fermi, E., 1, 1*n*, 4, 6, 7, 7*n*, 12, 12*n*, 16, 18
Feyerabend, P. K., 82, 83, 86, 88, 91, 203*n*, 217
Feyerabendian pluralism, 82
Finke, R. A., 144
Flach, P., 109
Fleck, L., 217
Fleming, A., 74, 180
Fodor, J. A., 142
Forrest, S., 193
foundational epistemology, *see* epistemology, foundational
Franklin, A., 65, 67
Franklin, N., 144
Fresnel, 117
Freud, S., 120
Frisch, O., 22

Gärdenfors, P., 104, 106, 108
Gabbay, D., 109
Galileo's dynamical theory
concepts of earth, water, and air, 29, 36–38
concepts of fall and natural place, 36–37
concepts of motion, 36–38
Galileo's method, 57
Galileo's theory of the tides, 37
Galileo, G., 169
and Copernicanism, 37, 37*n*
and the ship experiment, 37, 37*n*
and the tower argument, 37
Galison, P., 67
Gamow, G., 4, 7*n*
Garnham, A., 139
GAs, *see* genetic algorithms
Geertz, C., 135

gene expression, mechanism of, 46, 49, 50
gene regulation, mechanism of, 49, 88
gene replication, mechanism of, 49
gene segregation, mechanism of, 47
General Problem Solver, 175, 196
generatability, 173*n*, 176*n*, 184*n*
generation, *see* reasoning strategies, for generation
generativism, 174
generic abstraction, 148–149, 152, 153, 159, 160
generic modeling, *see* modeling, generic
genetic algorithms, 193, 193*n*, 194–196, 199, 200, 202
genetic programming, 194, 196–199, 201, 201*n*, 203
Genetic Programming Problem Solver, 196, 198
genetics
    developmental, 84, 86, 88, 89
    evolutionary, 86
    Mendelian, 44, 47, 49
Gentner, D., 129, 140
Gentner, D. R., 140
Ghiselin, M., 186*n*
Gibson, J. J., 133
Giedymin, J., 117, 119
Giere, R. N., 138, 153
Gilbert, S. F., 86
Gilhooly, K. J., 140
Glenberg, A. M., 144
Glennan, S. S., 45, 47
God design model, 185, 187, 189
Goldberg, D., 193*n*
Goldschmidt, R., 89
Gooding, D., 129, 151
Gould, S. J., 87, 89, 92, 182, 182*n*
GPPS, *see* Genetic Programming Problem Solver
GPS, *see* General Problem Solver
graded structures
    and acceptance, 4
    and similarity classes, 23
    and the discovery of nuclear fission, 4
    and the severeness of anomalies, 3–4, 23
    and the similarity account of concepts, 4
    individual differences in, 23
    psychological literature on, 4*n*
Greeno, J. G., 132, 133, 155
Griesemer, J. R., 138
Griffith, T. W., 140, 148
Gros, F., 53
Gruber, H., 75, 77
Györgyi, P., 234

H-creativity, 128–129, 212
Hacking, I., 66, 68
Hagstrom, W. O., 223, 224, 233
Hahn, O., 1, 2, 12, 20–23, 25
Hales, S., 72
Hall, N., 54
Hanson, N. R., 82, 83
Harré, R., 65
Harvey, W., 74
Hegarty, M., 145
Hegel, 167, 181–185
Heidegger, M., 69
Hempel, C. G., 109
heredity, mechanisms of, 47–49
Herrmann, G., 2
Hesse, M. B., 85
heuristic appraisal, 198
heuristic criteria
    analogy, 92
    and abduction, 83–86
    and causality, 84–85
    coherence, 83
    consilience, 86, 92
    consistency, 83
    explanatory potential, 62
    explanatory power, 83
    maximum likelihood, 84–86, 92
    precision, 83
    predictive power, 83
    prior plausibility, 84–85, 89, 91, 92
    problem-solving potential, 62
    truthlikeness, 83
    unifying potential, 62
heuristic pluralism, 82
heuristic procedures, 102
heuristic reliability, 83
heuristic rules, 175
hill climbing, 175, 202
Hoagland, M. B., 52
Holland, J. H., 140, 143, 193, 194, 198*n*, 202
Holmes, L., 54
Holyoak, H. J., 102
Hull, D., 177, 186*n*, 189, 197*n*
human design model, 183, 184, 186–189, 192*n*, 193
Hume, D., 186
Hutchins, E., 132, 135
Huygens, C., 169
hypothesis evaluation, 99
hypothesis generation
    and mechanisms, 44
    and visual thinking, 112
    constraints guiding, 44–45
hypothesis withdrawal, 111, 112, 114, 122*n*
hypothetico-deductive method
    and the BV+SR model, 202, 203
    in mathematics, 58
    Popper's version of, 177, 183*n*
hypothetico-deductive model of inquiry, 171*n*

ignorance
    and inquiry, 178

and luck, 178
imagery
and problem solving, 102–103
debate, 145
mental, 144, 145, 150, 157
incompleteness (of mechanisms), *see* failures (in mechanisms), incompleteness
inconsistencies
and abduction, 104–114
and auxiliary hypotheses, 113
and classical logic, 97, 183
and conceptual change, 110
and theory change, 97
handling, 111, 113
Popper on, 183
resolution of, 96, 106
inconsistency-tolerant logics, 183*n*
incorrectness (of mechanisms), *see* failures (in mechanisms), incorrectness
independent assortment, mechanism of, 47
induction
and abduction, 99
and deduction, 99
and diagnostic reasoning, 100
and hypothesis evaluation, 99
Baconian, 66
types of, 99
inference to the best explanation, 97, 99
innovation
and BV+SR, 176
and chance, 172, 178
and creativity, 210–211
and discovery, 175
and evolution, 182, 188–190
and knowledge-based systems, 176
and luck, 172, 178, 179, 179*n*
and prespecified goals, 191
and serendipity, 81
and supernatural faculties, 177
and tradition, 210–211, 213
and trial and error, 176
evolutionary model of, 184, 187
God design model of, 185, 187, 189
human design model of, 183, 184, 186–189, 192*n*
method of, 176, 178, 184, 203
theories of, 185
inspiration, 180
investigative pathways, 69, 71, 72, 75, 78
Ishida, Y., 167*n*
isotope, concept of an, 29, 36, 38, 39, 40*n*

Jablonka, E., 90
Jacob, F., 51–53, 88
James, W., 177
Johnson-Laird, P. N., 139–141, 144, 151, 156, 189
Joliot, F., 4
Josephson, J. R., 99
Josephson, S. G., 99
Judson, H. F., 49, 51, 52
Just, M. A., 145
justification
and discovery, 44, 63, 171, 174, 184
and scientific method, 169, 170
and thought-experiments, 60
consequential, 171*n*, 172
context of, 44, 184*n*, 190, 190*n*, 213
foundational, 202
generative, 171*n*, 172, 173*n*
logic of, 54, 171, 190, 190*n*
method of, 171, 201
justification, mathematical, 60, 63

Kaiser, M., 215
Kakas, A., 109
Kant, I., 34*n*, 119, 170*n*, 173
Kantorovich, A., 186*n*
Karrer, P., 225–227, 229–231, 233
Katsuno, H., 109*n*
Keilin, D., 229
Kelly, K., 191*n*
Kepler, 176*n*
Kepler's laws, 175*n*
Kimura, M., 182*n*
Kintsch, W., 140, 151
knowledge-based systems, 99, 176, 198
Konolige, K., 104
Kosslyn, S. M., 142, 144, 145, 156, 157
Koyré, A., 169*n*
Koza, J. R., 192*n*, 194–199, 201–203
Krafft, F., 2*n*
Krebs, H., 71–73, 76, 77
Kruse, R., 109
Kubler, G., 69
Kuhn, R., 225, 226, 229–231, 233, 234
Kuhn, T. S., 3, 34*n*, 57, 81, 175*n*, 203*n*, 211, 216–218, 235
Kuhnian monism
and consensus, 81
and scientific research, 81, 82

Lakatos, I., 57, 58, 63, 96, 96*n*, 113
Lakoff, G., 4*n*, 140
Lamb, M. J., 90
Langley, P., 175*n*, 209
language
and the world, 31
learning, 32, 136
Lanzola, G., 99
Latour, B., 74, 130, 138, 233
Laudan, L., 39*n*, 50, 173, 174, 174*n*
Lave, J., 132, 133
Lavoisier, A.-L., 71, 72, 77, 235
Le Grand, H. E., 65

Lea, G., 188
learning
 and BV+SR, 176, 197*n*
 method of, 176
 theories, 185
  instructionist, 186, 186*n*
  providential, 185, 186*n*
  selectionist, 186
Leibniz, 168, 169, 203
Levesque, H. J., 104
Levi, I., 104*n*
Lewontin, R., 92, 182, 182*n*
Liebig, J., 73
linguistics, cognitive, 128
Lloyd, B., 140
Lloyd, E., 212, 213
Lloyd, J. W., 115*n*
logic
 and scientific method, 170
 modal, 105
 non-classical, 97, 183*n*
 nonmonotonic, 105
 relevant, 97
 undecidability of, 117
logic of discovery, *see* discovery, logic of
logic of justification, *see* justification, logic of
Logic Theorist, 175, 188
logical empiricists, 174, 184*n*
Louis, S., 198*n*
Lovelace, Lady, 180
luck, *see also* chance
 and AI methods, 193
 and discovery, 168, 170*n*
 and ignorance, 178
 and innovation, 172, 178, 179, 179*n*
 and inquiry, 172, 178
 and method, 170, 172, 178–180
 methodological, 179, 180, 193
 moral, 179
 Popper on, 179
Lynch, M., 138

Machamer, P., 43, 45, 46, 54
Macready, W., 176
Magnani, L., 97, 99, 102, 104*n*, 112, 122
Malthus, T. R., 85
Mani, K., 139
Margolis, H., 189
Marx, 182, 183
mathematical growth, 64
mathematics
 argumentation in, 58
 concepts in, 30
 discovery in, 58
 hypothetico-deductive method in, 58
 justification in, 60, 63
 thought-experiments in, 57–64
Maxwell's method of analysis, 154
Maxwell, J. C., 85, 102, 103, 131, 146, 147, 149, 150, 153–155, 158–160
Mayo, D., 213
McCarthy, E. M., 92
McCollum Nickles, G., 167*n*
McDonald, J. F., 92
McNamara, T. P. , 139
mechanics
 Newtonian, 113
 principles of, 118, 118*n*
 quantum, 114
mechanism schemata, 45–47, 52, 53
 and anomalies, 54
 revision of incomplete, 47–50
 revision of incorrect, 50–53
mechanism sketches, 46, 47, 53
mechanisms
 analyses of the concept of, 45, 47
 and analogies, 91
 and anomalies, 50
 and diagrams, 44, 46, 50, 51
 and discovery, 43–46, 53
 and working entities, 46–50
 characterization of, 45–47
 discovery of, 45, 46, 49, 53, 54
 evaluation of, 44, 47
 evolutionary, 92
 examples of
  DNA replication, 46
  DNA synthesis, 49
  gene expression, 46, 49
  gene regulation, 49
  gene replication, 49
  gene segregation, 47
  heredity, 44, 47–49
  independent assortment, 47
  protein synthesis, 44, 46, 49, 50
 failures in, *see* failures (in mechanisms)
 in molecular biology, 46
 organization of, 45, 46, *see also* constraints on the organization of mechanisms
Meheus, J., 109, 167*n*, 183*n*
Meitner, L., 1, 7, 12, 20, 22
Mendelejeff's periodic law, 12, 24
Mendelian genetics, *see* genetics, Mendelian
Mendelzon, A., 109*n*
Meno, 178, 178*n*, 192*n*, 199*n*, 201*n*
mental imagery, *see* imagery, mental
mental modeling, *see* modeling, mental
mental models
 and external representations, 150
 and knowledge organization, 141
 and mental simulation, 155, 157
 and model-based reasoning, 152
 and problem solving, 155

and propositional representations, 143
and symbols, 143
and thought-experiments, 151–152
as working memory representations, 141
depictive, 144
dissemination of, 150
in cognitive science, 139
perceptual, 144, 157
mental simulation, 141, 144–145, 155, 157
Merton, R., 223, 232
Meselson, M., 53, 76
messenger RNA, discovery of, 52
metaphor, 200, 203
method
and BV+SR, 190, 193
and chance, 168, 178, 180
and creativity, 180
and luck, 178–180
of analysis, Maxwell's, 154
of conjectures and refutations, 177
of discovery, 168, 171, 175*n*, 176, 177, 181, 184, 198, 201, 203
of innovation, 178, 203
of inquiry, 169, 172
of justification, 171, 201
of physical analogy, 158
scientific, *see* scientific method
methodological criteria, *see* heuristic criteria
methodological luck, 179, 180, 193
methodology
modern, founders of, 175, 201
of discovery, 172, 178, 190
of science, 173
Popper's, 179
Mitchell, M., 193, 193*n*
model-based reasoning, *see* reasoning, model-based
modeling
analogical, 148, 152
and reasoning, 138
computational, 155
constructive, 104*n*
generic, 104*n*, 161
mental, 139–146, 151, 156, 158
practices, 138, 140, 145, 146, 148, 156, 158
visual, 149, 152–153
models
and analogs, 140, 141
and relation structures, 140, 142
external, 155
iconic, 143, 155
internal, 155
Maxwell's, 160
perceptual, 144, 157
physical, 140
propositional, 143
molecular biology, *see* biology, molecular
Monod, J., 51, 53, 88
moral luck, 179
Morange, M., 51
Morgan, T. H., 47
motion, *see* concepts, of motion
Moulines, U., 219
multiple discoveries, 220, 223

Nagel, T., 179
nature–nurture debate, 136
negation as failure, 97, 115*n*, 114–117, 121, 122*n*
neglect
of new discoveries, 3
of the nuclear fission hypothesis, 2, 23
Nersessian, N. J., 3*n*, 102, 104*n*, 128, 129, 138, 149, 159, 160
Nersessian, N. J., 109
networks
collaborative, 225
of enterprise, 77
neuroscience, cognitive, 128
neutron, discovery of the, 4
Newell, A., 129, 175, 175*n*, 196
Newton, I., 36, 66, 84, 127, 146, 149, 154, 168, 169, 173, 176*n*, 203
NFL theorems, *see* "No Free Lunch" theorems
Nickles, T., 167*n*, 169*n*, 171, 171*n*, 173, 174*n*, 176, 176*n*, 178*n*, 186*n*, 193*n*, 196*n*, 199*n*, 204*n*
Nisbett, R., 137
"No Free Lunch" theorems, 172, 176, 198
Noddack, I., 1, 2, 16, 18, 22–24, 24*n*
nominalism, 119
normal science, 211, 235
Norman, D. A., 132, 134, 155
Norton, J., 152
novelty, *see* innovation
nuclear disintegration, 4, 6, 20, *see also* disintegration processes
nuclear electron hypothesis, 4, 4*n*
nuclear fission
and Bohr's droplet analogy, 22
discovery of, 2, 2*n*, 4, 22
nuclear fission hypothesis
acceptance of the, 2, 23
neglect of the, 2, 23
raised by Hahn and Straßmann, 2
raised by Noddack, 2, 22
nuclear physics, 22
nucleus, conceptions of the, 4, 20

Oakhill, J., 139
Olby, R., 51, 52, 175*n*
oxygen, discovery of, 175*n*, 235

P-creativity, 128–129, 212

Paley, W., 185, 190
Pardee, A. B., 51, 52
Pauling, L., 233
Pearl, J., 105, 112
Peirce, C. S., 99, 101, 102, 167*n*, 183
Peng, I., 99
periodic table, 11, 12, 25, 39, 39*n*, 40
    anomalies in the, 38
    missing elements in the, 23
Perrig, W., 140, 151
Perry, R. B., 167
physical analogy, method of, 158
physical symbol systems, 129, 130
physics, nuclear, 4–22
Pickering, A., 177*n*
Plato, 169, 177, 199*n*
Plotkin, H., 177, 186*n*, 189
Pohl, R., 225
Poincaré, H., 114, 117, 118, 118*n*, 119–122
Poole, D., 106, 107
Pople, H., 99
Popper, K. R., 65, 66, 96, 101, 122, 174, 176–180, 183*n*, 184–186, 190, 190*n*, 199
predictive power, *see* heuristic criteria, predictive power
Priestley, J., 235
priority disputes, 223, 226, 229, 233, 234
problem solving
    and abduction, 82, 99
    and analogies, 102–103
    and concept formation, 102
    and conceptual change, 146
    and diagrams, 135
    and discovery, 43, 53, 175
    and distributed cognition, 134
    and explanatory promise, 82
    and external representations, 134, 155
    and imagery, 102–103
    and mental modeling, 146, 155
    and mental representations, 139
    and reasoning, 43
    and scientific research, 81
    and situated cognition, 133–134
    human design model of, 184, 193
    strategies, 44
progress, scientific, *see* scientific progress
protein synthesis, mechanism of, 44, 46, 49–53
Prout, W., 38
psychology
    cognitive, 102, 128, 139, 213
    neuro-, 156
Pylyshyn, Z., 142, 145

quantum mechanics, 114
quantum theory, discovery of, 175*n*
query evaluation, 114–117
Quine, W. V. O., 34*n*, 96, 109

radioactivity, discovery of, 39
Raff, R., 89
Ramoni, M., 99
reasoning
    abductive, 83
        formal models of, 104–109
        model-based, 107
        nonmonotonic character of, 100
    ampliative, 97
    analogical, 62, 88, 102
    and mental logic, 139
    and mental models, 140
    and modeling, 138
    and problem solving, 43
    case-based, 176, 198, 202, 203
    consistency-based, 104–109
    creative, 97, 99, 110, 112
    diagnostic, 50, 99–101, 108, 113
    distributed, 154–158
    imaginative, 156
    legal, 112
    manipulative, 104*n*
    medical, 99
    model-based, 104*n*, 146–153, 176, 202, 203
        and conceptual change, 137–153
        and distributed reasoning, 154–158
        and situated reasoning, 153–154
        and working memory, 141
        simulative, 156
    philosophical accounts of, 138
    productive forms of, 138
    situated, 153–154
    suppositional, 104*n*
    syntactical account of, 139
    traditional account of, 146
    visual, 102
    with incomplete information, 108, 113
    with inconsistent information, 108, 113
reasoning strategies
    for evaluation, 44–47, 53, 54
    for generation, 44–47, 52–54
        forward/backward chaining, 47
        modular subassembly, 47
        schema instantiation, 47
    for revision, 44–47, 50, 51, 53, 54
Reggia, I., 99
Reggia, J., 99
Reiter, R., 104, 106–108
Reitman, W., 167*n*
relation structures, 140, 142, 154
relativity theory, 113, 114
representation
    linguistic accounts of, 142
    perceptual accounts of, 142
representations
    amodal, 157

analog, 142, 143
and tokens, 143
external, 134, 136, 150, 154, 155, 157
iconic, 142–144
internal, 136
linguistic, 142
mental, 139
modal, 157
perceptual, 142
propositional, 142, 143
Rescher, N., 178
Resnick, L. B., 132
revision, *see* reasoning strategies, for revision
revolutions, scientific, *see* scientific revolutions
Rheinberger, H.-J., 49, 52, 54, 68–70, 72–74, 76, 78, 185*n*
Richardson, R. C., 45
Riley, M., 52
Rips, L., 155
Robotti, N., 38
Roentgen, W., 180
Rosch, E., 140
Roussel, P., 115
Roy, J., 54
Rudy, H., 225
Russell, B., 33

Samuel, A., 194, 196
Schaffer, S., 67, 175*n*, 220
Schaffner, K., 192*n*
Schelling, 167, 183
Schuster, J., 174
Schwartz, D. L., 144, 145
scientific growth, 109, 112
scientific method
and demarcation, 169
and discovery, 169
and essential definitions of science, 169
and justification, 169, 170
and logic, 170
and luck, 170, 172
and the diffusion of science, 169
and the explanation of discovery, 169, 170
and the progress of science, 169
and the Scientific Revolution, 169
and the unity of the sciences, 169
as a method of discovery, 171
as the One True Method, 171, 172, 174
attacks on the traditional view of, 171–181
characteristis of, 170
general, 170, 172, 174, 184
justification of, 172–173
modern, founders of, 168
traditional views of, 169–171
scientific methods
Baconian, 173–174
Cartesian, 173–174
deductive, 173–174
inductive, 173–174
self-corrective, 173
scientific progress, 70, 118, 169, 174, 179, 183*n*, 192
scientific research
and problem solving, 81
Feyerabendian strategy for, 82
Kuhnian strategy for, 81, 82
Scientific Revolution, 167, 168*n*, 169, 169*n*
scientific revolutions, 119, 211
Seaborg, G. T., 2
Segré, E., 4, 6
selective retention, *see* BV+SR
self-corrective methods, 173
Sellars, W., 29–36, 186
semantic view of theories, 210, 212–214, 218–221
serendipity, 81, 178, 179, 191, 209
set theory, classical, 33
Shanahan, M., 104
Shapin, S., 67
Shapiro, A. E., 66
Shelley, C. P., 111, 138
Shepard, R. N., 142, 144, 156, 158
Shepherdson, J. C., 115*n*
Shore, B., 132, 136, 140
similarity classes
and graded structures, 23
and taxonomies, 3
Simon, H. A., 129, 130, 142, 175, 175*n*, 188, 189*n*, 196, 196*n*
Simonton, D., 186*n*
situated reasoning, *see* reasoning, situated
Skipper, R. A., Jr., 45, 54
Slobodkin, L., 210, 215
Smith, A., 188
Smocovitis, V. B., 82
Sober, E., 84
social constructivism, and discovery, 175
sociology, cognitive, 128
Socrates, 178
Soddy, F., 39, 39*n*, 40
Solomon, R., 167*n*, 183
Souriau, P., 177, 178
Spirin, A. S., 51
Stefanelli, M., 99
Sternberg, R. J., 139
Straßmann, F., 2, 21–23, 25
Strong Programme in Sociology, 177*n*
Stuewer, R. H., 4*n*, 22*n*
Suchman, L. A., 132, 133
Suppes, P., 212
Swift, J., 190
symbol systems, physical, 129, 130
symbols
amodal, 143

and mental models, 143
modal, 143
synthetic a priori knowledge, 170, 173, 188

taxonomies
acceptance of changes in, 4, 23
and anomalies, 3, 4
and similarity classes, 3
Kuhn's theory of, 3
taxonomy
of chemical elements, 5, 6, 8, 22
of disintegration processes, 5–8, 12, 13, 16, 19, 20, 22–24
Teller, P., 193*n*
Thagard, P., 45, 47, 95, 96*n*, 99, 100, 102, 110–113, 129, 138, 213, 214
Theorell, H., 225, 229, 234
theory change
and abduction, 100
and anomalies, 111
and inconsistencies, 97
Thomson, W., 131, 159
thought-experiments
and Archimedes's method, 58–60
and argumentation, 63
and arguments, 152
and conceptual change, 151, 153
and conceptual problems, 57
and discovery, 60, 63
and Galileo's method, 57
and justification, 60
and Kuhn, 57
and mathematical discovery, 58
and mental imagery, 60
and mental models, 140, 151–152
and model-based reasoning, 151
and real experiments, 57, 58
and the discovery of the Euler formulas, 62
and the introduction of complex numbers in algebra, 60–62
in empirical science, 57
in mathematics, 57–64
Tomasello, M., 135, 136
Toulmin, S., 217, 218
transuranic elements
categorization of, 13
discovery of, 12, 16, 23
number 93, 1, 2, 12, 14, 16–18, 24
trial and error, 61, 173*n*, 176, 184, 189, 190, 192, 193
truthlikeness, *see* heuristic criteria, truthlikeness
Turing, A., 193
Tversky, B., 144
Tweney, R., 129

underdetermination, 202
unity of science, 169

van Assche, P., 2
van den Broek, A., 39*n*
Van der Waerden, B., 61, 62
Van Fraassen, B., 212
Vera, A., 129
Verbeurgt, K., 113
visual modeling, *see* modeling, visual
Volkin, K., 52
von Baer, E., 88
von Euler-Chelpin, H., 225, 227, 229, 230, 234
von Grosse, A., 1, 12, 12*n*, 14–16, 16*n*, 17, 22, 24
von Weizsäcker, C. F., 6, 20
Vygotsky, L., 136

Wagner-Jauregg, T., 230, 234
Wallace, A., 186, 197
Warburg, O., 72, 77, 224, 225, 229–231, 234
Wason, P. C., 140
Watson, J., 180, 233, 234
Watson, J. D., 49, 50, 53
Weart, S., 22
Weber, E, 183*n*
Weismann's rule, 91
Whewell, W., 172
whiggism, 169, 169*n*
Wiener, N., 185
Williams, B., 179
Williams, L. P., 159
Wilson, A. C., 87
Wimsatt, W., 45, 138
Winston, P. H., 167, 167*n*
Witten, E., 114
Wolpert, D., 176
Woods, D. D., 132, 135
Woods, J., 109
Woolgar, S., 74, 130, 138, 233
Wright, L., 183

Yeh, W., 157
Yeo, R., 174

Zamecnik, P., 69
Zhang, J., 134, 155

FSC
www.fsc.org
MIX
Papier | Fördert
gute Waldnutzung
FSC® C083411

Kolibri 360 GmbH
Ferdinand-Jühlke-Straße 7
99095 Erfurt, Deutschland
produktsicherheit@kolibri360.de